NOT A
SCIENTIST

W. W. NORTON & COMPANY

INDEPENDENT PUBLISHERS SINCE 1923

NEW YORK • LONDON

NOT A
SCIENTIST

HOW POLITICIANS
MISTAKE, MISREPRESENT,
AND UTTERLY
MANGLE SCIENCE

DAVE LEVITAN

For information about permission to reproduce selections
from this book, write to Permissions, W. W. Norton & Company, Inc.,
500 Fifth Avenue, New York, NY 10110

For information about special discounts for bulk purchases,
please contact W. W. Norton Special Sales at
specialsales@wwnorton.com or 800-233-4830

Manufacturing by Quad Graphics
Book design by Barbara M. Bachman
Production managers: Anna Oler and Lauren Abbate

LIBRARY OF CONGRESS CATALOGING-IN-PUBLICATION DATA

Names: Levitan, Dave.
Title: Not a scientist : how politicians mistake, misrepresent, and utterly
    mangle science / Dave Levitan.
Description: First edition. | New York : W.W. Norton & Company, [2017] |
    Includes bibliographical references and index.
Identifiers: LCCN 2016029048 | ISBN 9780393353327 (pbk.)
Subjects: LCSH: Science—Social aspects. | Technology—Social aspects. |
    Science—Political aspects. | Technology—Political aspects.
Classification: LCC Q175.5 .L4745 2017 | DDC 303.48/3—dc23 LC record
available at https://lccn.loc.gov/2016029048

W. W. Norton & Company, Inc.
500 Fifth Avenue, New York, N.Y. 10110
www.wwnorton.com

W. W. Norton & Company Ltd.
15 Carlisle Street, London W1D 3BS

1 2 3 4 5 6 7 8 9 0

To my wife, Jamie, and to Schrödinger's cat.
The cat will know why, but only if you ask it.

Contents

FOREWORD   ix

INTRODUCTION   1

1. THE OVERSIMPLIFICATION   9

2. THE CHERRY-PICK   28

3. THE BUTTER-UP AND UNDERCUT   44

4. THE DEMONIZER   60

5. THE BLAME THE BLOGGER   73

6. THE RIDICULE AND DISMISS   96

7. THE LITERAL NITPICK   111

8. THE CREDIT SNATCH   124

9. THE CERTAIN UNCERTAINTY   138

10. THE BLIND EYE TO FOLLOW-UP   155

11. THE LOST IN TRANSLATION   172

12. THE STRAIGHT-UP FABRICATION   186

CONCLUSION:
THE CONSPICUOUS SILENCE   201

NOTES   207

INDEX   241

Foreword

THROUGHOUT THE FOLLOWING PAGES, YOU WILL READ OF scientific errors and fabrications from a wide variety of politicians. You will see almost nothing, however, from President Donald Trump.

Do not let his absence from these pages fool you: what the new president does not know about science could fill a book on its own. The reasons for that absence relate not to Trump's actual statements on science, but in part to those statements' utter inanity, and in part to timing.

To the latter point first: I completed this book before Trump's victory in November 2016, and even well before his nomination began to seem likely as the primary season progressed. Almost no one, pundit or politician alike, foresaw his ascendance, and during the writing process he seemed to me and to many others little more than a sideshow. By the time that assessment proved laughably false, the window to significantly change the content you find here had long since passed, and thus the leader of the free world is mostly absent from a book on politics and science.

The other reason to largely ignore him lies in the specific ways President Trump errs on science. As you will find throughout this book, there are a variety of subtle, nuanced, and sometimes downright malevolent ways to manipulate science toward political ends. Many elected officials are quite skilled at these methods of deception, and debunking their errors requires as much care and attention to detail as their crafted talking points did in the first place.

Donald Trump is not like those elected officials.

In short, his errors on scientific topics are so blatant, so crude, so lacking in even the most basic understanding of physics or biology or chemistry or any other discipline that debunking them often requires essentially no effort at all.

"The concept of global warming was created by and for the Chinese in order to make U.S. manufacturing non-competitive," he tweeted in 2012. The preposterous nature of this claim, and many others like it, renders any sort of analysis unnecessary. It's just ridiculous; there is no evidence for it and in fact mountains— entire mountain *ranges*—of evidence to the contrary. And while some of Trump's errors could find a home in the last chapter of the book ("The Straight-Up Fabrication"), most are more effectively laughed off than analyzed.

Of course, this is a terrifying state of affairs in which to find ourselves. A president who believes vaccines cause autism (they don't, as you will see in several chapters), who thinks torture works (it doesn't, along with being morally repugnant), who thinks wind turbines are the primary killers of birds (in reality, not even remotely close) just because he doesn't want the turbines near his golf course—this doesn't bode well for many of the pressing science-related issues facing the country.

Further, while many of President Trump's errors are silly to an extreme extent, this does not mean that those surrounding him at the White House and in other leadership positions are so lacking in subtlety. With such anti-science attitudes now residing at the very top echelons of government, the tools presented here have arguably become more important. Holding our elected officials accountable for their misrepresentations of science could have remarkably far-reaching effects, with the impending doom related to climate change on top of the list.

As you read through these pages and learn to spot the various types of blunders and obfuscations, remember that President Trump's mistakes on science would be right at home in this collection, if only he was better at making them. Since he isn't, consider that he may have instead carved out his own brand of error: the FIREHOSE. Debunking an individual lie or error becomes much more difficult—even if it is an absurd error along the lines of "the Earth is flat"—when there are dozens, hundreds, thousands of errors coming at you all at once.

This is where the danger Trump poses to scientific progress and more generally to the nation becomes clearer; hitting one tennis ball back across the net is easier than hitting 50. Which balls will pass us by? A regulation on oil and gas drilling that ignores the danger to water supplies? An executive order that sea level rise should not be considered in federal projects? Or just a series of late-night tweets denouncing state-level vaccination mandates?

Vigilance is the only antidote against a flood of misinformation, deception, and backwardness. If you spot any particularly egregious misuses of science from the President or any other politicians, call your senator or House representative—let them know that you want Washington to curtail its anti-science ways. With the longer view in mind, support efforts to improve science, technology, engineering, and math education, through nonprofits like Change the Equation. Collectively standing up to and fighting back against anti-science governance will keep the march of progress on its steady path.

NOT A
SCIENTIST

Introduction

··

I N OCTOBER OF 1980, IN THE MIDST OF HIS FINAL CAMPAIGN
push against incumbent President Jimmy Carter, Ronald Reagan addressed some environmental concerns during an event in
Steubenville, Ohio. In his speech he presaged a trend in political rhetoric that would not truly rear its head for another three
decades:

> I have flown twice over Mount St. Helens out on our
> West Coast. *I'm not a scientist* and I don't know the figures, but I just have a suspicion that that one little mountain out there has probably released more sulfur dioxide
> into the atmosphere of the world than has been released
> in the last 10 years of automobile driving or things of
> that kind that people are so concerned about.[1]

It's true: the man who would soon become the fortieth president of the United States was not, in fact, a scientist. In spite of that lack of scientific training, he managed to express some opinions on science and science-related policy, in this case about how much sulfur dioxide—the primary culprit in the formation of acid rain—that "one little mountain" emits compared to human sources. Of course, Reagan also wasn't an economist, or an immigration lawyer, or an expert on the Middle East, but he had plenty to say and do about those issues as well. In general, the president doesn't need to remind us of the things he is not qualified to do; we tacitly trust that in spite of those shortcomings, he might have a few experts on hand to answer some questions.

As it turns out, Reagan's claim about Mount St. Helens, which had erupted in dramatic and deadly fashion months earlier, was quite far off. An Environmental Protection Agency representative at the time told the *New York Times* that although the volcano spewed as much as 2,000 tons of SO_2 per day on average, all human sources in the United States produced about 81,000 tons per day. Globally, at the time, the total would have been over 300,000 tons of sulfur dioxide from human sources each day.[2]

The massive eruption of Mount St. Helens alone released about 1.5 million tons, according to the US Geological Survey, which is certainly no small amount.[3] And if Reagan really meant to compare this to "10 years of automobile driving," he might have been closer—except automobiles are not the issue when it comes to SO_2. By far the biggest source is the burning of fossil fuels at power plants, accounting for 73 percent of the total today, followed by other industrial facilities—factories—at about 20 percent.[4] Ten years' worth of SO_2 emissions from "things of that kind that people are so concerned about," meanwhile, was equal to more than 200 million tons from the United States alone.

Okay, so Reagan was wrong. Who cares? After all, he *said* he wasn't a scientist. Can you blame him for being off on the details?

Of course we can. Simply saying you're not an expert is not an appropriate introduction for trying to act like one. In this example, the arguments over sulfur dioxide did not end in 1980; the Reagan administration obfuscated and delayed for years when it came to efforts to curtail SO_2 emissions and thus cut down on acid rain. Spreading misinformation like this— one volcano dwarfs human contributions—is a key component of keeping the public on your side, or at least keeping the public in the dark so that people can't object to your actions.

Fast-forward a few decades, and "I'm not a scientist" has become a go-to talking point for certain politicians, most often deployed in arguments about how to combat—or whether to combat—climate change. It is a dodge, a bit of down-home hucksterism designed to marginalize those eggheads over there who actually *are* scientists as somehow out of touch or silly. It is a simple way to get out of the conversation entirely, and it is often accompanied by bits of misinformation on the science as well.

Finding examples of its use is not difficult. In 2009, then-candidate for US senator from Florida Marco Rubio answered a question about whether climate change is caused by humans: "I'm not a scientist, I'm not qualified to make that decision. There's a significant scientific dispute about that."[5] (No, there is not.)

Florida governor Rick Scott, after years of outright denial

that humans had anything to do with a warming world, altered his position in 2014: "Well, I'm not a scientist." He went on to tout his administration's environmental record on specifics like "flooding around our coast"—without mentioning the relationship to rising seas and climate change—before saying it yet again: "But I'm not a scientist."[6]

Former Ohio congressman and Speaker of the House John Boehner added a mild wording wrinkle, saying in 2014: "Listen, I'm not qualified to debate the science over climate change."[7] He followed up by saying that any efforts to deal with the problem would wreck the economy—which most experts also agree is totally untrue.

Senate majority leader and Kentucky Republican Mitch McConnell sounded a familiar climate dog whistle when he brought up the 1970s "global cooling" red herring (in reality, very few were actually concerned about this, and global warming was already considered much more mainstream science): "We can debate this forever. George Will had a column in the last year or so pointing out that in the 70s, we were concerned the ice age was coming. I'm not a scientist."[8] The obvious question: McConnell thinks listening to George Will is wise, but listening to thousands of actual scientists around the world is somehow going to lead him astray?

There are other examples of politicians using the "not a scientist" line, or slight variations on it, all designed to pretend like the science is somehow unknowable. And of course, these politicians are not relinquishing their positions of power to those who *do* know about the topic in question. They go right on legislating and governing on the very topics they claim to know so little about; the most important piece of the talking point may be the "but" that follows it. The meaning of the line—that these individuals are literally not trained in science—has nothing at all to do with the questions at hand or what they were elected to do. The line makes so little sense that a GOP consultant and strategist, Mike McKenna, called it "the dumbest talking point in the history of mankind."[9]

The "not a scientist" line is a way out of talking about actual science, but politicians don't always have such an exit strategy when scientific topics arise. And as those examples illustrate, even when they do try to dodge the question, they end up spewing misinformation and errors virtually at every step. This book is about what happens when our elected officials talk science, and fail.

They do so in a wide variety of ways: sometimes there is subtlety and nuance; sometimes there is fiery or inflammatory rhetoric; sometimes they seem to snatch nonsense out of the ether. Yet their methods do follow some larger patterns that can be pinpointed. This book groups these rhetorical and logical errors into clear types to help you find them when they arise, and to cut through the misinformation once you've spotted them. You will see the Internet's nefarious influence with the BLAME THE BLOGGER, sniff out the unmistakable intent behind the BUTTER-UP AND UNDERCUT, learn to look for the bigger picture with the CHERRY-PICK, and glimpse the inner workings of many others.

This is not just to impress your friends during the next presidential debate, State of the Union address, or contentious legislative battle. Scientific topics have increasingly jumped to the fore of many political discussions in recent decades, and letting officials run wild with scientific missteps means there is less of a check on how those issues are actually legislated and regulated. Politicians have large platforms—increasingly so, with the advent of social media—and can reach wide audiences in short periods of time. A member of Congress spouting that vaccines cause autism is not without effect: the rise of the anti-vaccination movement, quite clearly a cause of recent outbreaks of relatively forgotten diseases such as measles, is a direct consequence of such incendiary and misinformed speech. And spreading misinformation about climate change has helped delay any global movement to stop it for decades, meaning that scientific ignorance or obfuscation on the part of our leaders could quite literally destroy the world as we know it.

Some important notes before we begin: First, with a couple of exceptions, there is little effort here to ascribe intent to the errors. It is very difficult to pinpoint exactly *why* politicians make mistakes when discussing science; maybe they truly don't understand a complex issue, or their ideological agenda makes them honestly skeptical of a given scientific topic. Or maybe they're just lying to gain votes, pass a bill, or garner support from a particular industry. Even when we get to the STRAIGHT-UP FABRICATION, teasing out reasons is difficult. The point is that intent doesn't matter: getting science wrong has a whole host of effects, from eroding public trust in both politics and the scientific community, to actively harming people today and in the future.

Second, some of the science described in this book will undoubtedly have evolved in various ways by the time it reaches readers' hands. In fact, the march of scientific progress is even the subject of a chapter: the BLIND EYE TO FOLLOW-UP. When writing about areas of active research, the best one can say is that it represents the available knowledge at the time.

Finally, the "not a scientist" line was originally a Republican talking point, and the vast bulk of errors (though not all) explored in this book were made by Republican politicians. This is not meant to be a partisan statement; it is merely a reflection of an unfortunate reality in recent years and decades—that one of the two dominant American political parties has largely abandoned mainstream scientific viewpoints. This should not be a particularly startling or controversial statement; indeed, entire books have been written specifically on the GOP's strained relationship with science.[10]

Some individuals within the party itself have decried the antiscience attitudes: former Louisiana governor Bobby Jindal, who has himself used a variation on the "I'm not a scientist" evasion, has also said that the GOP needs to "stop being the stupid party."[11] South Carolina senator Lindsey Graham issued this challenge: "To my friends on the right who deny the science, tell me why."[12]

The goal here is not to pile on one party further, but to highlight the types of errors we hear when listening to any political speech, from any party. The errors in recent years have come largely from the GOP, but that is not to say that the tables might turn at some point in the future. The techniques described here could be deployed by anyone, covering a wide variety of specific scientific topics, including those that haven't yet become an issue in today's politics. By laying out how these methods work, and why our leaders get these particular scientific issues wrong, we can recognize the errors when they surface in the future. Hopefully, the next generation of politicians will not be able to get away with them quite so easily or so often.

CHAPTER 1

The Oversimplification

On MAY 13, 2015, LAWMAKERS ON THE FLOOR OF THE HOUSE of Representatives in Washington, DC, took turns standing up and boiling down remarkably complex science into quick sound bites. Their topic: whether an unborn child feels pain at specific points of development while in the womb, and whether abortions should be banned after the earliest of these points to avoid any suffering for the fetus.

Here's Charles Boustany, from Louisiana:

> The scientific evidence is clear: unborn babies feel pain. They feel pain at 20 weeks post-fertilization.

Next up, Dan Benishek, from Michigan:

> The Pain-Capable Unborn Child Protection Act will prevent abortions from occurring after the point at which many scientific studies have demonstrated that children in the womb can actually feel pain.

And one more, Ralph Abraham, also from Louisiana:

> As a doctor, I know and I can attest that this bill is backed
> by scientific research showing that babies can feel pain at
> 20 weeks, if not before.[1]

Other legislators—all Republicans—made very similar statements. An average listener may have heard these statements
and thought, *Okay, it sounds like science has found the precise point at
which a fetus feels pain.* Interestingly, most of the Democrats who
stood up and spoke in opposition to the bill did so on grounds
unrelated to the science in question, but instead focused on the
bill's attack on women's rights and government control of one's
body. This could have been a good opportunity, though, to add
to that message and simply say: The science on this topic is *far*
from settled.

The problem with researching fetal pain is that pain is a subjective experience. This is why doctors will ask a patient: "On
a scale of 1 to 10, what is your pain?" There has been exten-
sive research into pain scales, pain tolerance, and related issues,
because pain is complicated. All we can say about pain is what
the person in pain can tell us about it, and of course an unborn
child is incapable of telling us whether something hurts. As a
result, the science on this topic is almost impossible to settle,
and we're certainly not there yet. But if you listened to all those
House members, you would think we did a few studies, got a
result, and hey, we're done! Fetal pain at 20 weeks is a certainty.

This is the most basic of the mistakes, misrepresentations, and
mangling of science that this book will address: the OVERSIM-
PLIFICATION. Unlike the science discussed, this idea is straight-

forward: politicians often make strong, definitive claims about topics that have far more subtlety than those statements allow or acknowledge. They take a complicated scientific issue and strip it down to a sound bite, a pithy turn of phrase that might garner cheers during a speech or build support for a piece of legislation, but would also give anyone listening an incorrect impression of the science in question.

This isn't to say that there is malicious intent behind any particular OVERSIMPLIFICATION. Sometimes, indeed, it is useful to take complicated issues and make them easier for the public to understand, even if it means leaving out a few details. But politicians aren't always just trying to help their constituents understand an issue. Sometimes they use the technique to obscure the truth, which—when it comes to science—can be messy, confusing, and unsettled.

So what *do* we know about fetal pain? Again, we can't ask a fetus whether anything hurts, so most of the research into this topic involves neuroanatomy and neurodevelopment. Basically, at what point are the connections completed between, say, our limbs and our brain, where the pain is actually processed and experienced? Though we can't really know just how a fetus perceives something that an adult would consider painful—like a pinprick, which a fetus can experience when a pregnant woman undergoes amniocentesis—we can at least know whether the signal gets from the site of the painful stimulus up to the brain.

And in fact, most published research suggests that the Republicans are wrong; the connections in question likely do *not* exist prior to about 23 or 24 weeks of gestation, if not even later. The connections needed for pain experience are between a part of the thalamus, which is a sort of relay center in the brain, and the cortex, the big part of the brain where many experts think pain is actually processed. Without connections from nerve endings to the spinal cord, from the spinal cord into the thalamus, and from the thalamus up into the cortex, pain simply may not be possible.

Several reviews of available evidence, conducted by well-respected scientific organizations, have concluded that the insistence on pain at 20 weeks is misplaced. For example, one 2010 synthesis of the evidence by the Royal College of Obstetricians and Gynaecologists in the United Kingdom concluded:

> In reviewing the neuroanatomical and physiological evidence in the fetus, it was apparent that connections from the periphery to the cortex are not intact before 24 weeks of gestation and, as most neuroscientists believe that the cortex is necessary for pain perception, it can be concluded that the fetus cannot experience pain in any sense prior to this gestation.[2]

The Texas branch of that group's US counterpart, the American Congress of Obstetricians and Gynecologists, agreed with that conclusion and specifically addressed the periodic attempts to pin down 20 weeks in legislation, writing in 2013:

> Supporters of fetal pain legislation only present studies which support the claim of fetal pain prior to the third trimester. When weighed together with other available information . . . [the] supporters' conclusion does not stand.[3]

And finally, a synthesis of available evidence was published in the prestigious *Journal of the American Medical Association* (known as *JAMA*) in 2005, by researchers at UC San Francisco. They concluded, as others have, that "evidence regarding the capacity for fetal pain is limited but indicates that fetal perception of pain is unlikely before the third trimester."[4] The third trimester begins at 27–28 weeks from conception.

The supporters of the 20-week ban point to a few sources of evidence, though they don't stand up to much scrutiny. In one commonly cited paper from the journal *Behavioral and Brain Sciences* in 2007, Swedish neuroscientist Bjorn Merker wrote about "evidence that children born missing virtually all of the cerebral cortex nonetheless experience pain."[5] In other words, maybe that fully formed cortex isn't even necessary to experience pain, which could push that developmental milestone earlier by a few weeks. Merker's paper, however, wasn't about fetuses in particular, and the children he wrote of were anencephalic infants (babies born without large parts of the brain, who generally do not survive long after birth)—not quite the same question. In fact, Merker himself told the *New York Times* in 2013 that his paper had only "marginal bearing" on the question of fetal pain.[6]

Another common argument goes back to amniocentesis—a procedure to test for certain birth defects or genetic issues, as well as infections, in which a needle is inserted into the amniotic sac. If the needle pricks the fetus's heel, the fetus may recoil— just as you might pull your hand back from a hot stove. That response certainly may *look* like the fetus is experiencing pain, but let's go back to the 2005 *JAMA* review to explain this phenomenon: "Flexion withdrawal from tactile stimuli is a noncortical spinal reflex exhibited by infants with anencephaly and by individuals in a persistent vegetative state who lack cortical function."[7] In other words, the body is capable of perceiving and responding to *harm*—a process called nociception—without the brain actually processing it as *pain*. A reflex can move your limbs even without your knowing it. Pain requires some degree of consciousness, while nociception requires no such thing.

This all may sound like a complicated jumble of conflicting research. Note the language used in the *JAMA* conclusion quoted earlier: "fetal perception of pain is *unlikely*." Even some of the world's experts on the topic, writing in a journal carefully reviewed by other experts, are unwilling to state conclusions in as concrete terms as the GOP House members did. The lawmakers used words like "clear" and "settled," aiming to close off any and all debate on the topic. This is a hallmark of the OVERSIMPLIFICATION: science is often far muddier than a politician is willing to admit.

Furthermore, those politicians don't enjoy being called on their loose use of complicated science. Again, boiling down a complicated issue to a pithy quote makes for good speechifying, but it also often backs the speechifier into a scientific corner:

"Thing X is true!"
"Okay, where is your evidence for thing X?"
"It's . . . complicated."

Congressman Ralph Abraham, one of the legislators quoted at the beginning of this chapter, gave a fine illustration of this problem, and of the general disconnect between politics and science, when asked for evidence to support his claims about fetal pain. His spokesman sent an e-mail that, instead of offering up any actual studies as evidence, claimed simply that Abraham's expertise should not be questioned, given his training as a doctor.

> [Congressman Abraham] has read countless medical journals and articles during that career that have led him to the conclusion that babies feel pain at 20 weeks. There's no single article or fact sheet that led him to this conclusion; he reached it during an entire career of study.[8]

This is a stunning misunderstanding of how science works, for a doctor or for a lawmaker. "Trust me, I'm a doctor" does not exactly prove one's point. If you heard it from your own doctor when asking about a particular medication or procedure, you might decide it's time to switch doctors. It was essentially a Jedi mind trick, a wild hope that "you don't need to see the evidence" would somehow convince people that they did not, in fact, need to see the evidence. Remember, Abraham said on the House floor that fetal pain at 20 weeks is "backed by scientific research." If that were the case, it shouldn't be hard to actually cite that research.

That's where the OVERSIMPLIFICATION tends to break down. When you start to ask for and examine the evidence behind a scientific claim, you often see that science doesn't quite cooperate with political sound bites.

A KEY STRATEGY for spotting this type of error is to look at the underlying policy position behind the science in question. With regard to fetal pain, the claim regarding 20 weeks is part of the ongoing effort by Republicans to limit and restrict abortions. When placed in that context, it's not difficult to understand why "settled" science on fetal pain could help the political cause, and why the more nuanced take on the issue would undermine the legislative effort.

Here's another example of how the OVERSIMPLIFICATION is used to serve a specific policy position: in April 2015, New Jersey governor and GOP presidential candidate Chris Christie appeared on a radio show and said definitively, "Marijuana is a gateway drug."[9]

That sounds bad! The term "gateway drug" is likely familiar to most people. What Christie was saying was, essentially, if you smoke pot, you're well on your way to a needle sticking out of your arm. He made this clear, definitive claim—it *is* a gateway drug, not that evidence suggests it might be, or that it could be considered, or anything of the sort—as a way to support a policy position: his belief that states that had recently legalized recreational marijuana should not be exempt from federal laws that still considered marijuana an illegal and dangerous drug. Christie said that, as president, he would still prosecute marijuana offenses, even in Washington and Colorado, where pot was legal at the time of his comments. Why? Because gateway drug, that's why.

But once again, this is a fairly drastic OVERSIMPLIFICATION. Scientifically speaking, it is not accurate to say that marijuana is a gateway drug. It would also be inaccurate to claim the opposite, that marijuana is definitively *not* a gateway drug. The science, once again, is complicated.

The gateway hypothesis essentially says that the use of one

substance can lead to—or at least increase the likelihood of—the use of another. If you ask hardcore antidrug folks—like Governor Christie, apparently—the gateway effect for marijuana is real and important. If you ask pro-legalization folks, of whom there are a growing number around the country, you'll find the opposite claim: marijuana is definitively *not* a gateway substance. The reality, as is often the case, lies somewhere in the middle.

"The scientific community is still arguing about it," said Susan Weiss, an expert with the National Institute on Drug Abuse (NIDA), which is part of the National Institutes of Health. "It's a really complicated thing to tease out. It has been very contentious over the years."[10]

Responsible scientists are not afraid to admit this lack of definitive conclusion. "We don't know the answer" is not an admission of defeat, but just an acknowledgment that science is hard, that biology and chemistry and physics don't always offer a single, easy answer. Politicians don't like that sort of uncertainty.

There are two distinct ways a gateway effect could occur: (1) biologically, meaning that using one drug actually changes your brain or body in ways that make use of another drug more likely; (2) socially or culturally, meaning that the context in which you use or abuse the first substance might make the second more readily available, or more likely to be used as well. Some evidence suggests that both of these mechanisms do function with marijuana, though there is also evidence that suggests otherwise, or that other legal drugs, including nicotine and alcohol, have similar or even more dramatic gateway effects.

In 1999, the Institute of Medicine, which is part of the National Academy of Sciences (and has since changed its name to the National Academy of Medicine), released a report entitled *Marijuana and Medicine*, which addressed the gateway hypothesis and laid out a remarkably common problem when it comes to science and health in particular:

> In the sense that marijuana use typically precedes rather than follows initiation into the use of other illicit drugs, it is indeed a gateway drug. However, it does not appear to be a gateway drug to the extent that it is the *cause* or even that it is the most significant predictor of serious drug abuse; that is, care must be taken not to attribute cause to association.[11]

Correlation does not equal causation! This truism should ring out in classrooms, in newsrooms, in the halls of Congress and the White House—anywhere that science is discussed and acted upon. Just because one circumstance is connected to another, or follows another, does not mean that one *caused* the other.

One famous example is particularly illustrative: when ice cream sales increase, so do violent crime and murder rates; therefore, delicious summer treats are to blame for homicides, right? Obviously not; this is just coincidence. There is ample evidence that violence peaks in summer months[12] (why that happens is another unanswered question), when ice cream happens to become a lot more desirable than in January. Ice cream does not, in fact, send all of us into a murderous rage.

With regard to a gateway effect, just because the use of drugs like cocaine or heroin or methamphetamines tends to follow chronologically behind the use of marijuana does not mean that use of that first drug *caused* use of the latter.

To a politician, that fundamental tenet of science can get in the way of a good narrative. Wanting to appear hard on drugs and on crime, Christie made use of the sequential nature of drug use to make pot seem worse than it is.

However, he's not 100 percent wrong: some evidence does actually support gateway effects of marijuana; it's just not definitive, as Weiss of NIDA said. Some studies, done in rats and other rodents, have made intriguing findings. For example, in one study published in the journal *Neuropsychopharmacology* in 2007, researchers treated adolescent rats with THC (tetrahydrocannabinol), the main active compound in marijuana.[13] Those rats, when they reached adulthood, were given a way to self-administer heroin. The rats treated with THC used more heroin than other rats that had not been treated with THC. Another study, in 2014, published in the journal *European Neuropsychopharmacology*, similarly found that THC exposure in adolescence seemed to engender lasting changes to the rat brains.[14] One more, published in 2004 in *Biological Psychiatry*, found that THC exposure increased rodent tolerance for other drugs, meaning it could increase the use of cocaine, morphine, and others later in life.[15]

These studies, though, were all done in animals, meaning extrapolation to humans is difficult. Furthermore, the findings are not at all unique to marijuana. Nicotine and alcohol have shown very similar effects in other animal studies—drugs that, of course, are perfectly legal.

There is some limited evidence of a gateway effect in humans, using studies of twins. For example, a study published in *JAMA* in 2003 followed sets of twins that were "discordant" for marijuana use: one twin had used the drug by the age of seventeen, and the other had not.[16] This is a way to tease out any genetic underpinnings for a given effect, since identical twins share a genetic code: if genetics is responsible *rather* than a gateway effect, the discordant twins should have similar rates of drug use later. But they didn't. In that study, the pot-user twin had a 2.1- to 5.2-times higher risk of using other drugs, becoming dependent on alcohol, and overall drug abuse or dependence than the pot-naïve twin. In other words, according to that study at least, genetics cannot explain the drug use, and thus the gateway effect gains support.

But that's not the whole story—not remotely, as it turns out. As Susan Weiss, of NIDA, said: "Did marijuana change that twin and make them more likely to use other drugs? What was it about that one twin that made them use marijuana while the other twin didn't? We don't know the answer to that. Did he happen to have friends that were more deviant? It's very difficult to completely interpret these things; most likely there is probably some convergence of factors."[17]

Indeed, other studies have questioned that twin study's conclusion. One published in *Development and Psychopathology* a few years later found the same link as the first—but only in nonidentical twins. This is remarkable: nonidentical (fraternal) twins don't share DNA the way identical twins do, meaning that contrary to the first study, genetics apparently *do* play a role in drug use patterns. Put another way, the genetically identical siblings seemed to follow similar drug use patterns regardless of early exposure, but those who had different genes seemed to experience a gateway effect from early use. The authors of this second twin study concluded that "the 'gateway effect' might be better conceptualized as a genetically influenced developmental trajectory."[18]

Would Christie's point have sounded as strong, as presidential even, if he had said: "Marijuana likely plays a role in a genetically influenced developmental trajectory, similar in fact to nicotine and alcohol, and therefore I will prosecute pot offenses even in states that have legalized it"? That's not as straightforward as, simply, pot is bad and you should go to jail if you use it.

All that conflicting evidence touches on only the biological mechanisms that could underlie a gateway effect; the cultural and social side comprises another set of research altogether.

The argument here is much easier to understand: if you are someone who uses marijuana habitually, you are likely going to be exposed to other types of drugs as well, just by being around other drug users and dealers. That concept actually makes a gateway effect hard to tease out—the direction from pot to, say, cocaine, likely reflects the easy access to marijuana in the United States as compared to the harder-to-find drugs like cocaine or heroin.

Some sociocultural studies have actually suggested a causal link between marijuana and harder drugs. One example is an ongoing study in New Zealand looking at 1,265 individuals born in 1977, whom researchers have followed for more than three decades, measuring various health and developmental parameters. The study's lead author, David Fergusson of the University of Otago, has said that at least some of the data from this cohort of people "clearly suggest the existence of some kind of causative association in which the use of cannabis increases the likelihood that the user will go on to use other illicit drugs."[19]

If that sounds convincing, don't get too excited. He went

on: "Where things get murky is in the area of the nature of the causal process." We think pot leads to other drugs! But we have no idea how or why.

The availability of a given drug certainly does seem to play a role. One 2010 study comparing drug use patterns across countries found that "early-sequence" drugs including marijuana, alcohol, and nicotine did predict the use of other drugs—but the availability of each of those early drugs made a difference as to the strength of the gateway effect. For example, Japan has very low rates of marijuana use—in that study, only 1.6 percent of people tried it by the age of twenty-nine—and it also has higher rates of harder illicit drug use *before* using those in the alcohol/pot/nicotine group. These findings suggest that limiting access to marijuana—which, of course, was the policy position that led to this lengthy we-don't-know-much-of-anything discussion—may not have much effect on limiting other drug use.

Since Governor Christie's point was that he would enforce federal marijuana laws even over the state laws that made it legal, here is another interesting factor to chew on: the illegal status of the drug may, in fact, contribute to its gateway effects.

This makes rational sense: accessing one illegal drug probably opens doors toward accessing others. In contrast, going into a store to buy some beer or a pack of cigarettes doesn't do much of anything to improve one's access to meth.

There is evidence that legal status does matter—from everyone's favorite legal-pot paradise, the Netherlands. In that country, 15 of every 100 cannabis users have tried cocaine at least once. That's a lower rate than in countries where marijuana remains illegal, including Scotland, Italy, and Norway.[20]

Again, the lesson here, and through much of this book, is simple: science is *hard*. When a politician makes it sound easy, settled, definitive—look closer.

LET'S EXAMINE ONE MORE example of the OVERSIMPLIFICA-TION. This example illustrates another danger when it comes to dumbing down—even slightly!—a complicated scientific topic. Here's President Barack Obama with a simple declarative statement:

2014 was the planet's warmest year on record.[21]

Obama said this a number of times over the course of 2015, as he launched an aggressive, multifaceted campaign aimed at staving off some of the worst effects of climate change. And it sounds convincing, definitive: of all the years for which we have temperature data, 2014 (at the time he said it; 2015 subsequently obliterated the record, and 2016 is likely to do so once more) was the absolute warmest of the bunch.

Simple, right? Well, actually, no. Sort of. Somewhat? This is likely the most subtle and, honestly, nitpicky claim that this book will address. The president was *almost* correct, but not entirely—and the other lesson of the OVERSIMPLIFICATION is that even that tiny bit of wrongness can leave a politician open to far more misleading attacks by his opponents.

Obama's warmest-year claim came from the agencies responsible for actually measuring the world's temperature: the National Oceanic and Atmospheric Administration (NOAA) and the National Aeronautics and Space Administration, better known as NASA. NOAA's monthly report on climate for December 2014 included this statement:

> The average combined global land and ocean surface temperature for January–December 2014 was the highest on record among all years in the 135-year period of record, at 0.69°C (1.24°F) above the 20th century average.[22]

Again, pretty definitive! But that wasn't the only publication NOAA and NASA offered regarding 2014's place in the record books. They also put together a joint presentation, which included this slide:[23]

RANKING OF RECORD YEARS IS SENSITIVE TO METHODOLOGY AND COVERAGE

NOAA

Probability of warmest year	
2014	~48%
2010	~18%
2005	~13%
2013	~6%
1998	~5%

NASA

Probability of warmest year	
2014	~38%
2010	~23%
2005	~17%
1998	~4%

January 2015 | NOAA/NASA – Annual Global Analysis for 2014 5

Credit: NOAA/NASA

Navigate past the bit of jargon at the top, and you can see 2014 atop a list of other years, with a "probability of warmest year" number next to it. What this means is that, according to NOAA's calculations, 2014 had approximately a 48 percent chance of being the warmest year ever recorded; NASA gave it even less of a chance, about 38 percent. Measuring average temperatures across the entire globe is complicated and comes with a degree of uncertainty; these percentages reflect that uncertainty. In fact, according to NOAA, that 1.24°F above the twentieth-century average mentioned in its December 2014 statement had a range of uncertainty of 0.16 degrees in either direction. So, it *could* have been only 1.08 degrees above the twentieth-century average, which might move it down the list of warm years.

Was Obama flat-out *wrong* about 2014's spot on the temperature leaderboard? Not quite. What those tables tell us is that 2014 was, at the time, more likely than *any other individual year* to have been the warmest ever. In fact, it was more than twice as likely, according to NOAA, as the second-most probable option, 2010, which clocked in at an 18 percent chance. But if the question was which is more likely to have been warmer—2014 or *any* other year (in other words, 2014 versus the field)—then the scales actually tip slightly in favor of the field, at 52 percent over 48 percent.

Nonetheless, Obama was just about as right as he could be, without quite getting there. Again, no other individual year had remotely as good a shot as 2014.

A NOAA climate scientist named Deke Arndt explained it to Andy Revkin of the *New York Times* this way:

This may seem pedantic, but it's an important point: there is a warmest year on record. One of the 135 years in that history is the warmest. 2014 is clearly, and by a very large margin, the most likely warmest year. Not only is its central estimate relatively distant from (warmer than) the prior record, but even accounting for known uncertainties, and their known shapes, it still emerges as easily the most likely warmest year on record.[24]

So President Obama was essentially correct, but somewhat imprecise—a largely forgivable version of the OVERSIMPLIFICATION type of error. Though forgivable, here's why it still matters: it gave climate change skeptics and deniers something to jump on. NOAA's 48 percent mark doesn't sound like "warmest ever" to the untrained ear, and NASA's calculation actually pinned the number a bit lower. Those percentages led to headlines like this, from the *Daily Mail* in the United Kingdom: "NASA Climate Scientists: We Said 2014 Was the Warmest Year on Record . . . but We're Only 38% Sure We Were Right."[25]

Clunkiness of headline writing aside, this sort of triumphant cackling from climate deniers only fed the confusion that many in the general public likely have regarding climate change and climate science. Many such articles came from a certain corner of the media, and one could sympathize with a confused reader who might simply throw up his hands and say: "The scientists don't even know what's happening to the climate. Why should we do anything?"

By adding "probably" to his talking point, the president might have staved off some of the criticism, but he also could have simply skipped the "warmest year" line and used only his follow-up talking point: "Now, one year doesn't make a trend, but this does: 14 of the 15 warmest years on record have all fallen in the first 15 years of this century."[26]

In NOAA's list of possible record years, you might have noticed that all were recent. This is the far more important point than simply which year squeaks out the statistical victory for warmest ever: the world has warmed dramatically in recent years, and regardless of which specific year wins out, *all* of the warmest years have come since 1998. After Obama's claim, 2015 beat out 2014 for warmest year again—by a remarkably large margin, with a *94 percent certainty*[27]—and 2016 is projected to reset the record yet again; though there will again be some degree of uncertainty, it will only further confirm the disturbing trend in rising temperatures.

As we've seen, the OVERSIMPLIFICATION can cheapen the magnificent complexity of science in pursuit of the perfect sound bite. Even just the examples in this chapter show that this technique has some very real impacts: abortion restrictions limit women's access to reproductive health services, misinformation on drugs could contribute to overcrowded prisons and ruined lives, and feeding the climate deniers can only hurt attempts to save the planet. Don't be seduced by simplified versions of science; they may sound convincing, but odds are they're not entirely true.

CHAPTER 2

The Cherry-Pick

THE UNITED STATES SENATE: A PLACE OF DECORUM AND history, of respect for centuries-old institutions, of heady debate and a solemn and weighty legislative process. And of half-assed snowball fights.

Senator James Inhofe, a Republican of Oklahoma, has long led the congressional climate denier caucus. This is an unofficial group, and one that over a decade or so went from the fringes to the mainstream of GOP legislators. Inhofe has made countless speeches claiming to expose the supposedly fraudulent world of climate science, helped kill bills that would have tried to put a dent in warming temperatures and rising seas, and even crashed international negotiations in an attempt to undermine progress.[1] And on February 26, 2015, he brought a snowball—an actual ball of snow—onto the floor of the Senate.

"You know what this is?" he asked, sly grin on his face. "It's a snowball. From outside here. So it's very, very cold out. Very unseasonable."[2] He then tossed the snowball to the Senate president and urged him to catch it.

Again, it was February, in Washington, DC, so cold is essentially the definition of "seasonable," but let's ignore that for a moment. Whether or not the cold and snow were season-appropriate, the idea of using a single data point—in this case, one cold day, or one snowstorm or snowball itself—to supposedly prove a larger point is among the most fundamental of logical fallacies. This is also the error that may sound most familiar: the CHERRY-PICK, when you selectively pull out only those pieces of information that suit you, ignoring the larger body of evidence.

As we'll see, the CHERRY-PICK takes a number of forms, and Inhofe's snowball is only one example, sometimes referred to as a fallacy of anecdotal evidence. Inhofe's argument was simple: it's cold out, so the people who say the world is warming can't possibly be correct. This is, of course, ridiculous. No climate scientist has ever said that global warming means there will be no more cold days, ever, in parts of the world where a season called "winter" happens. But it is a remarkably common refrain: every time a major snowstorm hits, or a cold snap freezes New York or Chicago for a few days, the deniers jump on their soapboxes and crow that global warming can't possibly be happening.

They stay conspicuously quiet in the face of competing evidence, these doubters or deniers; in fact, because controversy has reigned in the media over what exactly to call these folks ("skeptic" is too kind, "denier" too narrow, "doubter" too vague),[3] let's coin a new term: climate "TOADS," or Those who Oppose Action/Deniers/Skeptics. The TOADS don't just ignore a particularly hot day in January, or even a few straight days over 100 degrees in Boston in July; there is plenty of real, non-anecdotal evidence to send Inhofe and his snowball home. If the climate were generally stable (that is, if it were not warming up dramatically), one would expect a similar number of daily record low temperatures and record high temperatures. After all, a daily record is just an aberration really; unlike the longer-term trends we call "climate," the short-term "weather" is full of noise. The fact that one particular February 26 was very cold, or very warm, says nothing about the long-term trends.

In a warming world, though—the world we are, unfortunately, living in—the record highs would outpace the record lows. Basically, the deck would be stacked in favor of higher temperatures, meaning it's simply more likely to achieve an aberrantly high mark than it is to set an aberrantly low mark. And that's exactly what we observe.

During the first decade of the twenty-first century, record highs in the United States outpaced record lows by a ratio of 2.04 to 1, according to one study published in late 2009.[4] That means for every record low temperature (for example, this particular November 12 was the coldest November 12 we have ever recorded in this particular location), more than two record highs were set. That beats out the ratios from the 1990s (1.36 to 1), the 1980s (1.14 to 1), and so on.

More recent data have shown that this trend is continuing. From 2010 to 2015, record highs were still outpacing lows by about 2 to 1.[5] And that ratio is likely to skyrocket soon.

That same 2009 study projected what will happen with record temperatures over the rest of this century, and the results are striking. That 2-to-1 ratio, remarkable enough on its own, will be a thing of the past. By the middle of the century, record

THE CHERRY-PICK • 31

highs will beat out record lows by a ratio of *20* to 1. By 2100, every record low will be matched by *50* record highs.

Inhofe didn't even pick a particularly cold day to make his CHERRY-PICKED point. Even in a warming world, there will be plenty of winter days in Washington and elsewhere that, yes, have snow on the ground. Lobbing a snowball in the Senate doesn't undercut the broader trend in any way.

INHOFE'S VERSION OF THE CHERRY-PICK is a simple one: a single, limited experience used to try to make a far broader point. But there are other, more clever ways of deploying this technique.

In March 2015, Texas senator Ted Cruz broke out another favorite talking point of climate TOADS: the so-called warming hiatus. Climate scientists had observed a slowing in the rate of warming from the late 1990s through the early to mid-2010s, and this was dubbed a slowdown or "hiatus." Here's how Cruz discussed it in an interview with the *Texas Tribune*:

> The satellite data demonstrate that there has been no significant warming whatsoever for 17 years. Now that's a real problem for the global warming alarmists because all of the computer models on which this whole issue is based predicted significant warming, and yet the satellite data show it ain't happening.[6]

The first problem is that most people who cited this hiatus— Cruz included—actually described it incorrectly: it was a reduction in the *rate* of warming, not in warming itself. In other words, the planet got hotter slightly slower than it had managed in earlier years. But it still got hotter.

The second problem is that to claim that "no warming" happened, you have to engage in a serious bit of CHERRY-PICKING. It works like this: compare a particularly warm year from some time ago to a more recent year with a similar overall temperature, and you'll see an essentially flat line. In general, the TOADS use 1998, a year marked by an exceptionally strong El Niño, the weather pattern characterized by warm Pacific Ocean waters that can raise global temperatures. That year, the temperature "anomaly"—the globe's departure above or below the twentieth-century average—was +0.64 degrees Celsius (°C). In a more recent year, 2013, the anomaly was +0.66°C. If those were all the data points you had, wouldn't it seem like temperatures have been largely flat over that particular period? Almost no change at all. No warming to speak of, no reason to dramatically cut carbon emissions or anything of the sort!

Let's adjust the period we're examining by just *one* year at the beginning: in 1999, the anomaly was +0.42°C. Suddenly, we see a warming trend of about one-fifth of a degree, a large amount in the context of a global average. If we extend the data out to 2014, the jump gets even bigger—to a full third of one degree.

That's a dramatic difference! Suddenly, instead of no warming whatsoever, we have a huge increase in global temperature in just a few years.

But starting with 1999 is just as misleading. In both cases, we are CHERRY-PICKING start and end points for the series, when we should, in fact, be observing as long a trend as we possibly can. There will always be variation from year to year, just as there is variation in temperature, in precipitation, even in sea-level rise, from region to region. And the longer-term trend is unequivocal: the world has warmed by about 1°C over the last century. Here's the full chart from NASA, showing all the data we have as of early 2016:

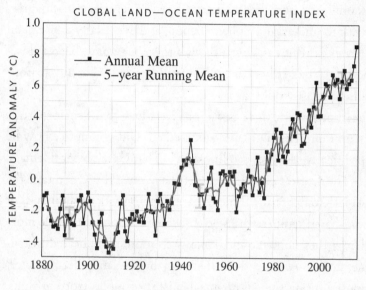

Credit: NASA Goddard Institute for Space Studies

Not only was the seventeen-year issue CHERRY-PICKED, but some new research after Cruz made his claims actually called the entire idea of the hiatus into question. Scientists with NOAA and elsewhere published a report in the prestigious journal *Science* in the summer of 2015 that recalculated global temperatures using the most up-to-date methods, and found that the idea of a hiatus was "no longer valid."[7] The warming trend from 2000 to 2014 was 0.116°C per decade, which was "virtually indistinguishable" from the longer-term trend from

1950 through 1999. The study's lead author told the Associated Press: "The reality is that there is no hiatus."[8] Of course, this revelation did not stop Cruz and other politicians from repeating the same claim throughout 2015, even after the *Science* paper was published. (Yet another paper in early 2016 found that there was, in fact, a slowdown in warming in the early 2000s[9]—an example of how honest and diligent scientists can disagree, work to understand a phenomenon, and eventually arrive at an answer. In this case, even though there may have been a slowdown, Cruz's silly claims went well beyond what any reasonable scientist would say.)

Impressively, there is still one *more* way that Senator Cruz's claim represented CHERRY-PICKED data. Did you notice how he specifically said the "satellite data"? That wasn't by accident.

The world's climate scientists have several ways to take the planet's temperature. One of them is a network of thousands of land-based temperature stations, some composed of little more than a thermometer and some with more complicated instrumentation. Another is with ocean buoys, and thermometers housed inside the engine mechanisms of globe-trotting ships. And yet another is a network of satellites, high above the Earth. The satellites use complicated instruments to estimate temperatures at various altitudes, including in the troposphere—the lowest part of the atmosphere, where we all live. (Satellites run by NOAA and NASA take many other important measurements related to climate, such as ice thickness, but we'll focus on temperatures here.)

All the different types of temperature measurements have different sources of error that require correction and adjustment. Skeptics have started homing in on the satellite data, because the various satellite temperature data sets don't paint as clear a picture as the other sources do. Specifically, one particular source of satellite data stubbornly doesn't show quite as much warming as the others. These data come from the University of Alabama at Huntsville.

The UAH record is managed by two professors, named John Christy and Roy Spencer. Both have ties to vocal climate-change-skeptic groups like the George C. Marshall Institute[10] and have published numerous opinion pieces in the *Wall Street Journal*[11] and elsewhere, downplaying the seriousness of the climate challenge and insisting that regulation of emissions isn't necessary. They are, quite clearly, outside the mainstream of accepted climate science.

Other climate scientists have shown that Christy and Spencer's data are not the most trustworthy, for various reasons. Here's how eminent climate scientist Raymond Pierrehumbert, now at Oxford and previously at the University of Chicago, described the situation for the well-respected *RealClimate* blog back in 2008:

> Spencer and Christy sat by for most of a decade allowing—indeed encouraging—the use of their data set as an icon for global warming skeptics. They committed serial errors in the data analysis, but insisted they were right and models and thermometers were wrong. They did little or nothing to root out possible sources of errors, and left it to others to clean up the mess.[12]

More recently, other climate scientists have applied some of the best available techniques to the same data that the UAH team uses, and have reached different conclusions.[13] Essentially, satellites—like any other source of measurement—aren't perfect. In a few different ways, they add slight bits of noise to the results; one of these ways is something called "diurnal drift."

That means that the satellite's orbit shifts slightly over its lifetime as it passes by overhead. In a study published in 2015 by University of Washington researchers, correcting for the diurnal drift in the UAH data set yielded a much stronger warming trend than Christy and Spencer report. This finding is borne out by the other satellite data sets, such as that from a company called Remote Sensing Systems and another by NOAA.

That the same raw data could yield differing outputs isn't particularly surprising. Experts say that the satellite data have more potential for error and should probably not carry as much weight as other sources. Even the scientists who themselves collect and analyze the satellite data say so! Carl Mears, a senior research scientist for Remote Sensing Systems, told the *Washington Post* the following:

> All datasets contain errors. In this case, I would trust the surface data a little more because the difference between the long term trends in the various surface datasets (NOAA, NASA GISS, HADCRUT, Berkeley, etc) are closer to each other than the long term trends from the different satellite datasets. This suggests that the satellite datasets contain more "structural uncertainty" than the surface dataset.[14]

By "structural uncertainty," Mears meant that we don't know as much about how the inputs—the actual readings taken by the satellites' instrumentation—are influenced by various factors. We can be more confident that the reading of a thermometer on the ground is telling us the truth (with some important exceptions, which we'll see in Chapter 5).

In a way, you can admire the thoroughness with which Senator Cruz crafted his talking point. He not only CHERRY-PICKED a starting year that enabled him to claim there has been no warming; he even CHERRY-PICKED the set of temperature records to armor himself against less-than-perfectly-informed criticism. Cruz seemed to love his satellite data idea: he mentioned it many times over the course of 2015, and even spent an extended period during a Senate committee hearing attempting to convince the president of the environmental group Sierra Club that no warming had occurred in seventeen years.[15]

But as we've seen, this isn't a particularly difficult claim to debunk. When a politician makes what sounds like a very specific point—no warming for *seventeen* years, not sixteen or eighteen—be wary. Check the details of that specificity. Did he pick that number for a reason? What do the longer, more general trends or more wide-ranging data show us? Where are the data actually coming from? If you ask these questions, this type of CHERRY-PICKING won't easily get past you.

THIS ERROR, THOUGH, is among the most expansive that this book will cover, with many types and versions available to a politician looking to throw a bit of murk over a scientific issue.

All these adaptations of the technique seem ready-made for the world of climate science.

When it comes to global warming, the problem is as broad as any that the scientific world encounters; it literally encompasses the entire planet. With countless data points on specific topics ranging from temperature to ice loss to extreme weather, a politician who wants to question the need for action can CHERRY-PICK to her heart's delight.

The ever-quotable Sarah Palin, former Alaska governor and 2008 GOP vice presidential nominee, used this tactic in criticizing President Obama's trip to her home state in the summer of 2015. Appearing on CNN's *State of the Union*, she suggested that Obama's visit to a receding glacier wasn't telling the whole story:

> I take changes in the weather, the cyclical changes that the globe has undergone for—since the beginning of time, I take it seriously, but I'm not going to blame these changes in the weather on man's footprint. Obama was up here looking at, say, the glaciers and pointing out a glacier that was receding. Well, there are other glaciers, though, that are growing up here. And he didn't highlight that, but he used glaciers as an example.[16]

Let's ignore her blithe dismissal of climate as "weather," her repetition of the common talking point regarding cyclical, since-time-immemorial change, her easily debunked idea that man isn't responsible (and it will be debunked in Chapter 9). Instead, we'll focus on the claim regarding glaciers—that "other glaciers . . . are growing up here."

As it happens, Palin was absolutely right about that; there certainly are glaciers that are gaining mass, even as temperatures warm. Palin used this point as a way to suggest that Obama is wrong to call for climate action: if *some* glaciers are actually gaining more ice, how could climate change possibly be a problem?

In accusing the president of CHERRY-PICKING his glacier, Palin actually was guilty of the CHERRY-PICK herself. She highlighted a certain glacier in Alaska, known as Hubbard Glacier, which has indeed been growing at a relatively steady pace since we began measuring it in 1895. Obama, meanwhile, visited Exit Glacier, which has receded a remarkable 1.25 miles over about two hun-

Exit Glacier, Alaska, which has receded more than
a mile since we began measuring it.

Credit: National Park Service/Fiona Ritter

dred years.[17] Obama called Exit "as good of a signpost as any when it comes to the impacts of climate change."[18]

Is this just a he-said-she-said kind of issue? They each have a glacier, one receding and one advancing, and neither is right about the more fundamental point?

Not even close. The fact is that the vast majority of glaciers both in Alaska and around the world are losing ice mass at remarkable rates. Some buck this trend and are gaining ice; that doesn't disprove global warming in any sense. In fact, in some cases it does the opposite, providing support for certain predictions about the effects of climate change. Here's how.

Glaciers, as you probably know, are essentially large blocks of ice. They gain mass, generally during winter months, through accumulation of snow on their surfaces. As pressure on the layers of snow increases, those layers gradually turn into ice. In the summer months, glaciers often lose mass, in the form of meltwater. If less snow falls in the winter than ice melts in the summer, the glacier has a "mass imbalance" and will shrink and recede. If the opposite is true—more snow falls on the glacier than it loses in ice melt in the summer—the glacier will expand, often "flowing" down mountains and valleys and cutting scars and holes in the terrain as it goes.

In a warming world, one would expect most glaciers to melt more than they gain in snowfall, and that's what we actually observe, as we'll see shortly. But some glaciers actually gain mass thanks to the warming climate: warmer air can hold more moisture, which means that in some parts of the world (remember there will be regional variation in climate effects; it isn't all one effect everywhere), precipitation rates will increase. If that precipitation takes the form of increased snowfall in an area where temperatures aren't high enough to cause summer melts that outweigh the new mass, glaciers will expand.

That's what's happening to the Hubbard: because its "catchment basin"—the area in which snowfall will contribute to the glacier's mass—extends over an enormous area, increases in precipitation in the region give the glacier a huge source of ice.[19] Its front edge has been pushing forward at a rate of between 13 and 36 meters every year for more than a hundred years.[20]

So, Palin picked a glacier that is likely expanding *because* of global warming to suggest that global warming isn't happening. Nice CHERRY-PICK! The more fundamental point, though, is that even if some are expanding, the vast majority of glaciers are shrinking.

A recent study on glaciers specifically in Alaska actually pinned a number on the ice loss: 75 gigatons per year. A gigaton, as silly as it might sound, is a no-joke amount of ice: 1 billion metric tons. If you're trying to picture how much this is, imagine 100 million African elephants, or 6 million blue whales, or *3,000 whole Empire State Buildings*.[21]

One scientist described a gigaton to the *Washington Post* in stark, visually arresting terms: take the National Mall in Washington, DC, from the Capitol steps down to the Lincoln Memorial (about 2 miles long and five football fields wide), and cover it in a giant block of ice four times as high as the Washington Monument.[22]

Have you managed to get a grip on that visualization? Now take seventy-five of those giant blocks of ice, chop them off of Alaska's thousands of glaciers, and dump them into the ocean. Once you've done that, we can start thinking about the rest of the world's glaciers as well.

Yes, it's not just Alaska, of course. The vast majority of glaciers around the entire globe, from the Alps to the Greenland ice sheet, are shrinking in this rapidly warming world. The World Glacier Monitoring Service, a United Nations program that keeps tabs on a number of "reference" glaciers, has found in recent years that between 80 and 90 percent of them are losing mass.[23] The National Snow and Ice Data Center at the University of Colorado at Boulder has said that over 90 percent of all alpine glaciers—which excludes ice sheets but includes the bulk of other glaciers around the world—are also retreating and losing mass.[24] The Greenland ice sheet itself has lost more than *9,000 gigatons*, or 9 trillion tons, since 1900.[25]

Even if you forget that warming temperatures are actively helping some glaciers gain mass, those stark numbers reveal Palin's claim as the CHERRY-PICK that it is. To the less well informed, though, she may sound like she has a point: Shouldn't warming temperatures make *every* glacier lose ice? And if one isn't, then why should I believe any of this climate change nonsense?

When a politician cites a singular example as a means of refuting a larger trend, take note; this is the hallmark of bad science. Ask yourself if that one counterexample really refutes the entire idea, or if there's more to the science than our leaders might say. Indeed, Obama's claim regarding Exit Glacier deserves the exact same scrutiny—but once scrutinized, it becomes clear that his example really is representative of the broader trend.

As we've seen, there are a number of ways to use the CHERRY-PICK that have subtle differences. You can take only a single data point, like the glacier example; you can zoom in on a particular section of a much bigger graph, as in the "seventeen years of no warming" claim; or you can focus on only a certain data *source* when the other sources disagree with your basic point. In any of these cases, the end result is the same: a small piece of a big puzzle is used to confuse anyone listening, and to undermine scientific and political progress at once.

....

CHAPTER 3

The Butter-Up
and Undercut

Some aspects of science and scientific research
are, from a public opinion perspective, unassailable. These tend
to be feel-good ideas—like, say, going to the moon—or abstrac-
tions that are simply impossible to argue with; who would try to
claim that research into how cancer forms is a *bad* idea?

But universal appeal doesn't mean that politicians are in
lockstep support of those issues. It just means they have to be a
bit more careful when they criticize them.

Here's an example, once again turning to Texas senator
and 2016 GOP presidential candidate Ted Cruz: during a Senate hearing regarding NASA's funding in March 2015, he lav-
ished praise on the space agency's history, its employees, and
its important position as a government entity.[1] He said that
"innovation . . . has been integral to the mission of NASA"
and spoke about the "passion of the professionals at this fine
institution." He quoted a former astronaut, saying that "young
Americans are interested in space-related STEM [science, tech-

THE BUTTER-UP AND UNDERCUT · 45

nology, engineering, and math] careers, and see themselves as future space entrepreneurs." He closed his brief statement by paraphrasing previous remarks about NASA: "It is time once again for man to leave the safety of the harbor and further explore the deep, uncharted waters of deep space."

And then he tried to cut NASA's funding for studying climate change.

See, Cruz wasn't there just to heap praise at NASA's doorstep and then heartily approve the budget requests that came in from the agency and from the White House. He was there to advance his own and the GOP's agenda as it pertains to climate change—an agenda that can be boiled down to "let's do absolutely nothing." NASA, though much more commonly associated with Mars rovers and moon landings, blue dots and Saturn's rings, is actually among the most important groups in the world for understanding how the Earth's climate is changing. It sends up satellites, monitors surface and ocean temperatures, flies missions over the Greenland and Antarctic ice sheets to measure melting, and conducts a variety of other scientific functions in the field. Without NASA's efforts, the field of climate science would be years behind where it is today.

Though climate science remains a controversial topic in certain political circles, NASA itself enjoys broad public and political support. In fact, a Pew Research Center survey in early 2015 found a 68 percent favorability rating for NASA, compared with only 17 percent unfavorable. For a government agency, that's remarkable; only the Centers for Disease Control and Prevention scored better, showing that the public really does appreciate science in a variety of forms.[2] (Yes, the IRS came in last.)

The many satellites and instruments that NASA uses
to measure climate, oceans, cryosphere, and more.

Credit: NASA

Given the generally positive opinion the public has about
our space agency, Cruz knew he had to be careful if he was trying to cut its funding. So he used the BUTTER-UP AND UNDERCUT,
a technique that equates to a magician convincing you to look
at the sparkly spectacle over on this side of the stage while he
makes the elephant on the other side "disappear." Politicians
looking to make what could be an unpopular point deploy this
technique as a way of distracting us—"Oh, Cruz loves NASA!
I agree, how cool was the Pluto flyby thing?"—while they get
to the messy work of undermining science. This ploy is often
used in funding battles, as we'll see; it's easier to pull the money
rug out from under an agency if it seems like you're an admiring superfan of that very institution.

Senator Cruz's version of the BUTTER-UP AND UNDERCUT
also happened to be full of errors about NASA's mission and

scientific foundations. Here's a longer quote from the hearing, in which Cruz claimed that the focus on our own planet was somehow at odds with the agency's overarching goals:

> Should NASA focus primarily inwards, or outwards beyond lower Earth orbit? Since the end of the last administration we have seen a disproportionate increase in the amount of federal funds that have been allocated to the earth science program at the expense of and in comparison to exploration and space operations, planetary science, heliophysics and astrophysics, which I believe are all rooted in exploration and should be central to the core mission of NASA. . . . I am concerned that NASA in the current environment has lost its full focus on that core mission.[3]

All of that is essentially code for "I don't want NASA studying climate change anymore." The use of "inwards" as opposed to "outwards," the focus on the concept of a "core mission"— these choices in phrasing are all intended again to highlight the grandiosity of NASA's space exploration, the wonder we all feel at a close-up of Jupiter, the cliché of a child wanting to grow up to be an astronaut. But Cruz couldn't have been more wrong.

Leaving the funding questions aside for a moment, it actually isn't difficult to find out what NASA's core mission really is. The space agency has a rich history that is almost entirely available online; deciphering its scientific function, in terms of both original intent and evolving mission, is just a matter of doing some reading.

NASA was created by the National Aeronautics and Space Act of 1958. This legislation was part of a hasty attempt to

respond to the Soviet launch of Sputnik, the world's first orbit-
ing satellite, in October 1957. The founding document actually
manages to refute Cruz in the very first objective listed: "The
expansion of human knowledge of phenomena in the atmosphere and space."[4]

The atmosphere! That would be *our* atmosphere, here on
Earth—the one you're living in and breathing and from which
you're drawing protection against the sun's rays. It's the atmosphere that has a temperature, which is warming up dramatically thanks to the burning of fossil fuels, a phenomenon we
know more about thanks to—NASA.

The act doesn't stop there. Another objective reads: "The
preservation of the role of the United States as a leader in aero-
nautical and space science and technology and in the application
thereof to the conduct of peaceful activities within and outside
the atmosphere." Again, it was clear from the very first document in NASA's history that this agency was designed not simply to send probes to the outer reaches of the solar system, but
to advance the study of our own planet as well.

This wasn't just foundational rhetoric either. It is true that
the early years of NASA focused strongly on getting humans
into space, and to the moon, but even then the agency's leaders
understood that studying our planet was of the utmost importance as well. Here's a document from 1964, three years after
President Kennedy set the goal of getting to the moon, describing the nonlunar aspirations of NASA:

> The fundamental objective of the Geophysics and Astron-
> omy Program is to increase our knowledge and under-
> standing of the space environment of the Earth, the Sun

and its relationships to the Earth, the geodetic properties of the Earth, and the fundamental physical nature of the Universe.

Knowledge of these areas is basic, not only to our understanding of the problems of survival and navigation in space, but also to the improvement of our ability to make technological advances in other fields.[5]

A decade later, with Neil Armstrong's "giant leap" a few years in the rearview mirror, NASA's budget estimates listed this as the very first of its achievements: "NASA's programs . . . extend man's knowledge of the earth, its environment, the solar system, and the universe."[6]

In 1984, as awareness of the realities of climate change and other global changes began seeping into the scientific community's consciousness, Congress actually revised the Space Act to include a line about "the expansion of human knowledge of the Earth."[7] In the 1990 budget request, NASA administrators set a series of goals and led off with: "Advance our scientific knowledge of Earth and of the forces and systems that shape our planet."[8] How would *you* define "core mission"?

Even a cursory reading of NASA history shows that studying our planet—in particular, its atmosphere—has always been central to the agency's existence. But if you listened to Cruz, you might have been blinded by the shiny spaceships over there and missed the attempt to disappear the elephant over on this side of the room.

Cruz's BUTTER-UP AND UNDERCUT had another couple of layers to it. Ostensibly, his point was that only a limited amount of federal money can go to NASA, and those useless Earth and atmospheric sciences are sapping the coffers of the *real* goals, of space exploration. As we've seen, Cruz was spouting nonsense in claiming that the agency had somehow shifted away from space and toward our own planet, but NASA administrator (and former astronaut) Charles Bolden pointed out some additional flaws in his argument:

> We can't go anywhere if the Kennedy Space Center goes under water and we don't know it. That's understanding our environment. . . . It is absolutely critical that we understand Earth's environment, because this is the only place that we have to live.[9]

Though this may sound like a neat rhetorical trick on Bolden's part, this is *not* exaggeration!

The Kennedy Space Center, where many space shuttle and rocket launches have taken place, sits on the Florida coast at Cape Canaveral. This part of Florida—well, okay, basically *all* of Florida—is extremely low-lying. Most of the area used by NASA is at or just above sea level. That means the warming climate, and the rising sea because of that warming climate, are a crucially important issue for the future of manned space travel.

In fact, some researchers have tried to quantify this issue. In a study published in the *Bulletin of the American Meteorological Society* in 2014, NASA and Columbia University experts found that the sea has been rising at Cape Canaveral at a pace of 22.6 millimeters per decade.[10] That pace, in Florida and around the world, is almost certainly accelerating[11] and will continue to accelerate in the next few years and decades. What does that

A rocket lifts off from NASA's Kennedy Space Center in Cape
Canaveral, Florida. Much of the center sits at or just above sea level.

Credit: NASA/Kim Shiflett

mean for NASA's primary rocket launch site? Floods. Lots
of floods.

Coastal flooding at Kennedy Space Center already happens
about once every ten years. By 2050, a conservative estimate is
that those floods will happen every three to five years. Before
too long, if we can't slow down the warming climate and the
rising sea, that site (and others like it) will be essentially unus-
able. Cruz wants to pull money away from climate science, but
that means spending a whole lot of money later: according to
the Columbia and NASA researchers, Kennedy's infrastructure
"would cost more to replace than at any other NASA site."[12]
Whoops!

This is another hallmark of the BUTTER-UP AND UNDERCUT:

the misdirection tends to weaken and damage the very institution or issue the speaker is admiring so breathlessly. Cruz's version builds up NASA's ego admirably, only to try to literally wash away its accomplishments.

There was even one more way Cruz erred during the hearing (yes, he managed to layer mistake on mistake, building an entire seawall of misleading statements, in just a few minutes of a generally positive-seeming speech). Here's Cruz again:

> That in my view is disproportionate, and it is not consistent with the reason so many talented young scientists have joined NASA. And so it's my hope that this committee will work in a bipartisan manner to help refocus those priorities where they should be, to get back to the hard sciences, to get back to space, to focus on what makes NASA special.[13]

Once again, NASA is "special"; its scientists are young and "talented." But what about that little remark about the hard sciences? Is Cruz suggesting that NASA is spending too much effort and money on "soft sciences"?

This is yet another branch of the BUTTER-UP AND UNDERCUT; the implication is that NASA's special status arises from its focus on "hard" scientific exploits, like journeys to Mars or the study of the sun. But the agency's earth sciences missions are a far cry from "soft" science.

The concept of hard and soft sciences isn't exactly written in stone, but there is general agreement that "hard science" refers to disciplines including chemistry, biology, physics, or astronomy—say, anything you might use a telescope or a microscope for. Soft science, on the other hand, includes fields such as psychology or anthropology—certainly worthy areas of study for thousands of researchers around the world, though clearly they aren't NASA's general cup of tea.

Cruz was trying to isolate space exploration and astrophysics as the only denizens of the hard-sciences landscape, shuffling things like oceanography or atmospheric science to the land of the psychologist. Quite clearly, that's a ridiculous distinction: NASA's measurements of ice sheets or ocean temperatures or forest cover fall under fields that are obviously hard sciences. Even the head of the august American Geophysical Union, an association of more than sixty thousand Earth and space scientists, took issue with Cruz's pooh-poohing of NASA's work. In a letter addressed to the senator, the AGU's executive director, Christine McEntee, wrote:

> Earth sciences are a fundamental part of science. They constitute hard sciences that help us understand the world we live in and provide a basis for knowledge and understanding of natural hazards, weather forecasting, air quality, and water availability, among other concerns.[14]

She pointed out that NASA's earth science priorities are based on reports known as decadal surveys, produced by the National Academy of Sciences. These offer a thorough examination of where the scientific priorities should lie, based on input from a wide variety of the country's best scientists. The last such survey, published in 2007, suggested that the "U.S. government, working in concert with the private sector, academe, the public, and its international partners, should renew its investment in Earth-observing systems and restore its leadership in Earth science and applications."[15] The funding changes that Cruz is so passionately against are a direct result of NASA—horrors!—*listening* to scientific recommendations.

Once again, it is clear that this type of rhetorical device can be particularly devastating, as it serves to mangle the foundational nature of the topic in question. NASA is great, sure, but some of its most important research is just about useless, if you take Cruz at his word. And NASA is not the only scientific institution to get buttered up on its way to a brutal teardown.

REMEMBER BIRD FLU? Avian influenza has had a few moments in the media sun in the last decade or two, sandwiched around some swine flu hysteria. The virus spread through birds of various types—first in Asia, and later in Europe and North America. Occasionally it managed to infect humans, and it had a high mortality rate when it did, but it proved to be generally difficult to transmit between people. In other words, bird flu represents one of many near misses when it comes to the world's next pandemic.[16]

Scientists have long warned that influenza in one form or another is likely at some point to pose a serious risk to humanity. The Spanish flu outbreak of 1918 killed more than twenty million people, and subsequent outbreaks in 1957 and 1968 killed many as well. Flu viruses change every year, mutate in subtle ways, and jump between animals of various types and people; it is a difficult threat to get hold of. And so, when avian influenza escaped the confines of Thailand and Cambodia and was found in birds in Turkey, Romania, Croatia, and even the

United Kingdom in October 2005, the world took it very, very seriously.[17]

So seriously, in fact, that President George W. Bush started throwing money at the problem. Bush visited the National Institutes of Health in Bethesda, Maryland, in November 2005 to discuss the threat of an influenza pandemic, to advocate for a massive funding push for research and preparedness, and to praise the great work that NIH scientists do. Skepticism is the appropriate response at this point.

Yes, Bush engaged in an example of the BUTTER-UP AND UNDERCUT at the NIH, though his effort was a bit more subtle than Ted Cruz's elephant-vanishing NASA trick. Here's part of Bush's speech to the NIH:

> For more than a century, the NIH has been at the forefront of this country's efforts to prevent, detect and treat disease. And I appreciate the good work you're doing here. This is an important facility, it's an important complex. The people who work here are really important to the security of this nation.[18]

The NIH is wonderful! Bush went on to outline specific goals and strategies for preventing and responding to pandemic disease outbreaks; he said he had requested $251 million from Congress to help train and equip foreign partners, $2.2 billion to purchase flu vaccines and antiviral medications such as Tamiflu, and $2.8 billion more for research into novel ways to produce those vaccines more quickly after an outbreak hits. These were good, proactive ideas, supported by most rational people. And if you listened to the entire speech, there would be no indication at all that Bush was anything but supportive of the important work done by scientists at the NIH and other institutions that receive funding from the NIH.

In this case, the misdirection was wholesale: Bush said one thing, but he spent years doing another. The NIH saw its total budget double in the late 1990s and early 2000s, which meant that scientists around the country were able to fund more projects and do more basic research on everything from Alzheimer's to diabetes. But once that doubling was complete, Bush's apparent disdain for science led to a stagnation in funding over several years; in fact, the NIH budget has actually declined since 2003, if inflation is taken into account.

When Bush gave his bird flu speech to the NIH, featuring his new requests for funding, fiscal year 2006 had just begun. The baseline NIH budget for that year was $28.56 billion, amazingly representing a decrease from 2005, when it was $28.59 billion.[19] A year later, the budget was nudged up to just $29.61 billion—an increase that barely kept up with inflation.

During these ongoing years of stagnation, scientists around the country have expressed dismay that basic science research is being shortchanged in favor of other budgeting priorities—like, say, a couple of wars. To be clear, President Obama has obviously played a role in budgeting during this period as well, and though he has not been working with a particularly friendly Congress, he is not free from blame. Toward the end of his presidency, however, Obama has managed to inject some large increases into the NIH coffers.[20]

Here's why this matters: the NIH is the primary source of funding for basic science research in the United States. There are other government agencies that fund research, such as the National Science Foundation (NSF), and individual universities certainly contribute some of their own money, but in short, as the NIH goes, so goes science in America. Scientists at every university and institute in the country submit grant applications—over fifty-one thousand of them in 2014 alone[21]—in the hopes that panels of experts will select their projects as worthy of funding. If you get funded, great, you're all set to bear down and get to work. If not? Well, hopefully your university can prop you up with some funding for a while, or you can find sources of money elsewhere. But you may find your job at risk if you can't secure an NIH or NSF grant within a few years.

There will always be some researchers whose work isn't quite up to snuff; we can't fund *every*thing. But expanding the NIH budget means being able to cast a wider net, to do basic research on a whole host of topics that, though they may sound silly when described literally (as we'll see in Chapter 6), could yield really important discoveries later on. So, keeping the NIH and NSF funded to a reasonable level is a crucial part of science and health. And if you listened to just President Bush's speech about bird flu, it would sound like he agreed with that sentiment.

Bush, though, clearly didn't agree. His NIH budget requests from 2003 through the end of his presidency seemed to say: "Meh." It's not like he proposed cutting the NIH completely or anything so drastic; again, some things, like basic research into diseases that affect all of us, are unassailable. But most people won't notice an ostensible drop in funding for basic science research, especially if you make speeches like Bush did, touting how the NIH and its people are "really important to the security of this nation."

He loves the NIH! Pay no attention to the cuts in funding that sent grant approval rates spiraling toward the single digits!

Bush's version of the BUTTER-UP AND UNDERCUT is something of a long con. In public, he said all the right things about how crucial basic research is to society. But when it came down to it, he didn't care enough about science to put the government's money where his mouth was. This duplicity played out over almost his entire time in the Oval Office. In 2007, after several years of stagnant NIH funding, Bush vetoed a spending bill that would have added an extra $1 billion to NIH coffers[22]— still not even enough to match the rate of inflation. In a speech, he likened Congress to "a teenager with a new credit card."[23]

The lack of support for basic research has had real, demonstrable effects. When the Ebola outbreak hit West Africa in 2014, and fear spread that this deadly disease could proliferate into Europe and North America, scientists rushed to try to find a vaccine. Francis Collins, the director of the NIH, had this to say about that effort:

> Frankly, if we had not gone through our 10-year slide in research support, we probably would have had a vaccine in time for this that would've gone through clinical trials and would have been ready.[24]

Again, basic science research often isn't the sexiest line item on a government ledger. Talking up the importance of science while cutting off the legs of the research community, though, ends up hurting all of us.

The BUTTER-UP AND UNDERCUT is among the more nefarious of the errors and rhetorical devices explored in this book. It carries an unmistakable air of intent: politicians have to *try* to use this technique, have to understand that they are walking a tightrope balanced between positive public opinion and negative action. By fixing our attention away from where the sneaky stuff is going on, they apply a magician's showmanship to the act of undermining scientific progress—an ugly bit of sleight of hand.

CHAPTER 4

The Demonizer

THE ERRORS WE'VE SEEN SO FAR HAVE INVOLVED SOME degree of subtlety. With the DEMONIZER, subtlety is chucked out the window. This rhetorical maneuver simply takes advantage of a difficult and usually scary scientific concept—often, the spread of dangerous diseases—and links it to an unrelated issue to advance a political agenda. Politicians most often deploy this tactic in efforts to criticize and curtail immigration.

For example, in early 2015 a disease straight out of the past made national news when a measles outbreak began at Disneyland. The outbreak was a direct result of the profoundly unscientific anti-vaccination movement,[1] but that didn't stop some politicians from pinning the blame on that eternal bogeyman, the foreigner. Here's an extended, grammatically challenged rant from Alabama congressman Mo Brooks, speaking on Matt Murphy's radio show:

I don't think there is any health care professional who has examined the facts who could honestly say that Americans have not died because the diseases brought into America by illegal aliens who are not properly health care screened, as lawful immigrants are. It might be the enterovirus that has a heavy presence in Central and South America that has caused deaths of American children over the past six to nine months, it might be this measles outbreak—there are any number of things. . . . unfortunately our kids just aren't prepared for a lot of the diseases that come in and are borne by illegal aliens. You have to have sympathy for the plight of the illegal aliens, I think we all understand that. But they have not been blessed with—in their home countries—with the kind of health care, the kind of immunizations that we demand of our children in the United States.[2]

This is an impressive example of fearmongering and contains a litany of scientific errors. Diseases are frightening things, especially those that sound unfamiliar to us—like the enterovirus Brooks mentioned (and was wrong about, as we'll see), or measles, which used to be common in the United States but was declared eliminated by the Centers for Disease Control and Prevention in 2000[3] (with sharp declines in prevalence decades before that). Brooks took advantage of that fear to connect the diseases to another issue that may be unpopular with his particular constituency: illegal immigration. In doing so, he spread decisively false scientific and medical information.

First, Brooks mentioned enterovirus. He was correct that this disease had caused the deaths of children in the United States, but he was wrong about its origins. The specific strain that caused problems in the United States in 2014 was enterovirus D68, which causes "mild to severe respiratory illness," according to the CDC.[4] Various other enteroviruses circulate in the United States every year, but the 2014 outbreak was a particularly severe strain. During that outbreak, assorted media personalities tried to pin the blame on a recent flow of undocumented immigrant children from Honduras, Guatemala, and El Salvador. Brooks was parroting that claim, and he spoke as though the connection had been firmly established. It had not.

In fact, no one knows exactly why there was a spike in D68 cases in 2014, or exactly where the virus came from. However, the CDC has said that "children arriving at U.S. borders pose little risk of spreading infectious diseases to the general public."[5] In the case of the enterovirus, certain studies also indicate that the flow of immigrants was unlikely to be the cause. One such study examined people across Latin America with flu-like symptoms and found that only 3 percent of them carried *any* enterovirus strain. Of that 3 percent, only 10 percent (meaning, 0.3 percent of the total, or three in one thousand individuals) carried the D68 variety.[6] In other words, the flow of children crossing the border almost certainly was *not* the source of the D68 outbreak in the United States.

Brooks, and the others who made the claim before him, had no evidence whatsoever that undocumented immigrant children were bringing this disease into the country with them. But criticizing undocumented immigrants in certain circles is a no-lose proposition, and Brooks took advantage.

Since his comments pertained to measles, it turns out Brooks was even more wrong. He suggested that all these children coming in from Central America don't have the levels of protection against disease that the United States has—that we are better at vaccinating against measles than the developing countries south of our border. That sounds logical; after all, the United States is among the most advanced nations in the world. Wouldn't it have better rates of measles vaccination than El Salvador?

To understand the answer to that question, a bit of background is useful. In 1998, a group of researchers led by UK physician Andrew Wakefield published a paper in the prestigious journal the *Lancet* with this impenetrable title: "Ileal-Lymphoid-Nodular Hyperplasia, Non-specific Colitis, and Pervasive Developmental Disorder in Children."[7] The paper argued, essentially, that the MMR vaccine—measles, mumps, and rubella—could cause autism in young children.

To be extremely, urgently clear: that conclusion was false. Vaccines. Do. Not. Cause. Autism.

Numerous studies have since looked for a connection between the MMR vaccine and autism, and have found nothing. It took a while, but eventually the *Lancet* retracted the paper entirely (in 2010) after finding severe flaws, while an investigation by the *British Medical Journal* turned up evidence of outright fraud.[8] Wakefield's medical license was stripped. Even before that, though, the Wakefield paper's influence was a bit bizarre, since the results it reported came not from a large cohort study of the sort that would ordinarily be required to find such a potential hazard with the MMR vaccine, but from only a small, observational study of a handful of children.

In general, the study and the still-resounding furor about vaccines and autism is perhaps the best-ever example of that golden rule: correlation does not equal causation. Autism spectrum disorders are generally diagnosed at an age range when children have recently received various vaccines, including the MMR; it is understandable that confused and worried parents look for something to blame, and the vaccine may have been the most recent connection to the medical world for that child. But in no way does that mean the vaccine *caused* the disorder.

Unfortunately, the long delay in retracting the study enabled the anti-vaccination movement to grow and metastasize, with the Internet furthering the cause by helping spread an entire genre of totally false and misleading vaccine information.[9] This movement led many parents to refuse vaccines for their children entirely, or at least to delay them, both of which increase the risk of disease not only for the unvaccinated children, but also for those around them. That trend, to withhold lifesaving vaccinations from children, has, against all odds, dropped the United States behind the developing countries that Brooks mentioned when it comes to protection against measles.

Some of the countries in Latin America that immigration hawks are so concerned about have similar or even better measles vaccination rates than the United States. According to the World Health Organization, the American rate of coverage among one-year-olds has ranged between 91 and 93 percent over the last couple of decades,[10] and certain small pockets of the country have seen those rates drop dramatically. Some places, such as Orange County, California, home of Disneyland and the origin of the 2015 measles outbreak, have seen huge drops in vaccination rates even as the national rate has stayed relatively high.[11] Across all of California, in 2000 only 0.77 percent of children started kindergarten with a "personal belief exemption" from vaccinations;[12] in 2013, that rate was four times as high, though it dropped a bit in 2015. And that statewide average masks the clumps of unvaccinated kids: some counties had exemption rates above 10 percent, and overall vaccination rates can drop into the low 80 percent or high 70 percent range.[13]

Those percentages are important because of a result of vaccination known as "herd immunity." Some children cannot receive vaccines for medical reasons; they may have a compromised immune system, say, thanks to leukemia or other diseases that require immunosuppressing therapies. Herd immunity is what protects those children from diseases like measles; it basically means that, with a high enough immunization rate, the disease can't take hold in a community and won't be able to infect vulnerable people. At the low levels of immunization seen in some California counties, herd immunity is severely compromised. Some parents argue that vaccinating their kids should be up to them alone, but the decision has actual, demonstrable effects on other people; even if you homeschool your unvaccinated children, are you going to keep them out of every other public space where they may interact with other children? Like, say, Disneyland?

The anti-vaccination trend, which we'll explore again in a later chapter, is, in large part, an American creation. Still, certain countries from which undocumented immigrants arrive do, in fact, have lower vaccination rates than the United States. For example, Guatemala's rate dropped to 85 percent in 2013, followed by a sharp dip to 67 percent in 2014. For a decade prior to that, though, the country had rates similar to those of the United States—at or above 90 percent (and the children who crossed the US border would likely have been of vaccination age during the period of higher vaccination rates).

Other neighboring countries do even better. Nicaragua's

measles immunization rate rose to 96 percent in 2005 and has held steady at an impressive 99 percent ever since. El Salvador's hasn't been quite as steady or as high, but in 2013 and 2014 it held at 94 percent. Mexico's rate in 2014 was 97 percent. The implication from Brooks was that these backward, disease-ridden countries were dumping their illnesses on rich, healthy Americans, but in fact the opposite is true. Coming to the United States from El Salvador might *increase* one's risk of disease, thanks to the lower vaccination rates here than back home.

The troubling vaccination trends in the United States, and the impressive public health immunization efforts in Latin America, didn't seem to matter to Brooks and others. Here's Ben Carson in 2015, at the time gearing up for his presidential bid, making a similar point:

> We have to account for the fact that we now have people coming into the country, sometimes undocumented people, who perhaps have diseases that we had under control. So now we need to be doubly vigilant about making sure that we immunize our people to keep them from getting diseases that once were under control.[14]

That's *Doctor* Ben Carson, by the way. Just as with Congressman Brooks, Dr. Carson seizes on the assumption that the United States is more medically advanced than the developing world. He was less obvious in his demonization of the "other," the diseased foreigner, but the implication is the same: we need to immunize "our people" because the other people might bring diseases with them.

It is not hard to draw a straight line from this type of scientific misstep to the anti-immigration policies espoused by many GOP politicians. Donald Trump's idea to build a (wholly impractical and prohibitively expensive[15]) border wall is in some ways a direct reaction to the idea of the diseased foreigner (obviously, economic fears and other factors play in as well). And of course, those types of fears are manifested in how the public votes. In other words, this misrepresentation of science and medicine can help produce actual policy that is unsupported by the scientific concepts behind it.

MANY OF THE OTHER ERRORS described in this book are relatively recent phenomena. The DEMONIZER, though, is perhaps the most persistent misuse of science that politicians have engaged in across US history. The fear of immigrants has been a common theme since the early days of Ellis Island—and even before—and there are examples of misstatements about disease and immigration from various points since.

Moving backward through history, we can see that politicians have repeated, time and again, the idea that immigrants are bringing over every disease that happens to be in the news. Here's former presidential adviser, 1992 and 1996 presidential candidate, and noted racist Pat Buchanan going for the DEMONIZER grab bag:

> High among these is the appearance among us of diseases that never before afflicted us and the sudden reappearance of contagious diseases that researchers and doctors

eradicated long ago. Malaria, polio, hepatitis, tuberculosis, and such rarities of the Third World as dengue fever,
Chagas disease, and leprosy are surfacing here.[16]

Leprosy! This is an incredible list of misinformation, so let's
debunk it disease by disease. First of all, malaria: between one
and two thousand cases of malaria are reported in the United
States every year, almost all in residents who have traveled
abroad to endemic areas. It is very unlikely to then pass from
person to person; a mosquito in the United States would have to
bite the infected person and then bite other people, transmitting
the parasite that causes malaria. Such outbreaks do occur—it
happened sixty-three times between 1957 and 2014, according
to the CDC—but this has nothing to do with immigration, but
rather with travelers failing to take basic precautions.[17]

Next up, polio. No cases of polio have originated inside the
United States since 1979, and the last time a traveler brought it
in was 1993.[18] That'll do.

It almost isn't worth addressing "hepatitis," as that refers
to five different diseases with differing modes of transmission.
Some evidence does suggest that prevalence of hepatitis B,
which is transmitted via bodily fluids, is higher among foreign-
born individuals living in the United States,[19] so we'll let him
have that one. Tuberculosis also may be brought across borders
by those arriving both legally and illegally, though this isn't
considered a large concern.

Next, what about those "rarities of the Third World"? Most
dengue fever cases are acquired outside the United States by
travelers, though interestingly, experts are concerned that climate change could allow dengue's spread northward by creat-

ing more habitat where the mosquitoes that transmit it can live. Fewer than forty cases of Chagas' disease—caused by a parasite that is transmitted by the so-called kissing bug—have been reported in the United States since 1955.[20]

And finally, leprosy. Yes, this disease still exists. Also called Hansen's disease, leprosy affected 2,323 people in the United States between 1994 and 2011, and Buchanan would be right if he noted it is more common among foreign-born individuals. But the idea that there is some explosion of this disease thanks to immigration is laughable; the rate of new diagnoses in that time period actually fell by 17 percent.[21] And here's another fun fact: some of the cases of leprosy in recent years have been transmitted not by people, but by armadillos.[22]

Perhaps it isn't surprising that Pat Buchanan's rhetoric on health and immigration isn't exactly scientifically sound. Going back a bit further, here's former Oklahoma senator Don Nickles during a 1993 Senate debate over trying to prevent HIV-positive foreigners from immigrating:

There are 700,000 immigrants that come into the United States every year. If we change this policy, it will almost be like an invitation for many people who carry this dreadful, deadly disease to come into the country because we do have quality health care in this country, better health care in the United States than any other country in [the world]. . . . I mention this amendment is not born out of hate. This amendment is not born out of fear. This amendment is not born out of homophobia. This amendment is raised to try and stop President Clinton's administration from making a very serious mistake that will jeopardize the lives of countless Americans and will cost U.S. taxpayers millions of dollars.[23]

The policy in question—preventing immigration of HIV-positive individuals—had been in place since 1987 (created by Senator Jesse Helms). The "countless" Americans, though, were a figment of Nickles's imagination; it was estimated that in 1989, for example, fewer than a thousand HIV-positive immigrants would even seek entry into the United States. Given HIV's limited modes of transmission, this was far from a public health crisis. Importantly, though, the DEMONIZER really works: the policy Nickles spoke about remained in place all the way until 2009, when President Obama finally lifted the ban.[24]

We can continue to go back in time. In 1915, for example, a typhoid fever epidemic began in Mexico,[25] spreading fear that it would penetrate the United States as well.[26] Along with a growing nativist sentiment and concerns about immigrants from elsewhere in the world, this epidemic led to passage of the Immigration Act of 1917. That ignominious bit of legislation lumped in the presumably diseased "other" with a whole host of supposed undesirables rivaling the Blazing Saddles army of "mugs, pugs, thugs," and so on:

The following classes of aliens shall be excluded from admission into the United States: All idiots, imbeciles, feeble-minded persons, epileptics, insane persons; persons who have had one or more attacks of insanity at any time previously; persons of constitutional psychopathic inferiority; persons with chronic alcoholism; paupers; professional beggars; vagrants; *persons afflicted with tuber-*

*culosis in any form or with a loathsome or dangerous contagious
disease.*[27]

Of course, the fears that some politicians took advantage of
at various points did at least have some basis in truth: there *were*
occasional disease outbreaks in other countries less advanced
than the United States, if one goes back far enough. But the
language of the DEMONIZER persisted long after many parts of
the world had begun to change. In a 2002 paper on the "persistent association" of foreignness and disease in the United
States, University of Michigan researchers Howard Markel and
Alexandra Minna Stern wrote about how much of the developing world modernized and left the American rhetoric behind:

> After World War II, many countries built hospitals and
> rural clinics and spearheaded campaigns to combat
> endemic diseases, and many parts of the world benefited
> from reductions in childhood mortality and various
> infectious diseases as well as improved standards of nutri-
> tion as a result of hygiene and maternity programs. In
> addition, organizations like the United States Peace
> Corps and the United Nations World Health Organiza-
> tion brought modern sanitary techniques, public health
> administration, vaccines, and medical treatments to areas
> that had neither the financial or human resources to
> afford them.[28]

And yet, half a century and more later, politicians continue to
connect foreigners with disease, even when those foreigners may
have grown up with better health care than the politician did.

Again, the DEMONIZER is an easy tactic for politicians to use, since the diseases in question can be scary and most people won't know that they are exceptionally rare in immigrants, or that vaccination rates are actually better in other parts of the world. Spotting this tactic is relatively easy, since it is generally limited to this particular scientific field: if a politician warns that allowing foreigners in will spread a certain disease, doubt the claim. Check the actual modes of transmission of the disease, or the actual prevalence of that disease. The devil isn't the immigrant; it's in the details.

....

CHAPTER 5

The Blame
the Blogger

THE INTERNET HAS BEEN DESCRIBED AS A GREAT EQUALIZER. Everyone with a connection can have a voice, whereas in the past the words of only a select few trickled out to the masses. In many ways, this is indeed a wonderful thing. In some ways— YouTube comment sections, say—it is not.

Just as with other topics, when it comes to science the Internet's essentially level playing field is a mixed bag. If you have a question about science, chances are there are resources online that can answer it, to the extent humanity is able to provide an answer. But chances are just as good that some source online can answer it incorrectly. For every Mayo Clinic, Science.gov, or NOVA website, there is a Mercola.com or Infowars muddying the waters. (Do not visit these last two—ever.)

Just as for the rest of us, picking out the good science from the bad can be tough for elected officials. And even if politicians know that a certain source may not be the best, sometimes they're not afraid to use it anyway. This is the BLAME THE BLOGGER—

when a politician repeats information from often terrifically dubious sources, with the knowledge that many people simply won't know how to check the underlying science. It's online, so it must be true!

In one example of this tactic from early 2015, an Alabama congressman named Gary Palmer went on a radio show and said some very wrong things about climate change and climate science. The host, Matt Murphy, set Palmer up expertly by asking him in traditional (for the TOADS, anyway) guffawing fashion his opinion on the relationship between snowstorms in the Northeast and climate change. He responded:

> I think it might be a matter of the report that came out last week about the government manipulating data and misleading people a little bit. But two feet of snow ought to get their attention. . . . We are building an entire agenda on falsified data that will have an enormous impact on the economy.[1]

A report that the government is "manipulating" and "falsifying" data? Well, that sounds horrific! It was not remotely true, of course, but Palmer didn't pull this out of absolutely nowhere; the Internet told him!

In fact, the claim of temperature data manipulation (which has popped up a number of times over the years) in this case originated with a blogger and climate denier named Paul Homewood. There was no "report," no official document or peer-reviewed research. It was just, to be crude, some guy online. (There is very little information about Homewood available on his website or elsewhere; one article about his work called him

THE BLAME THE BLOGGER · 75

a retired accountant,[2] and there is no indication that he has any expertise in climate science.)

Homewood wrote several blog posts with titles such as "Massive Tampering with Temperatures in South America."[3] His blog wasn't exactly CNN's home page in terms of traffic, so that incendiary headline might have remained obscure if it hadn't been for a TOAD with a bigger audience, Christopher Booker, writing for the *Telegraph*, a London newspaper. Booker covered Homewood's work and even managed to one-up him in the headline battle: "The Fiddling with Temperature Data Is the Biggest Science Scandal Ever."[4] Ever! Booker's stories were shared hundreds of thousands of times, and seized upon by TOADS everywhere.

Well, okay, what scandalous fiddling had Homewood and Booker brought to the world's attention? As it turns out, the process these shrill doomsayers had uncovered could easily be labeled "science." More specifically, Booker described a well-studied, well-understood process regarding temperature measurements—one that was not hidden from the public or from other scientists in any way, shape, or form.

See, taking the world's temperature is exceptionally complicated. There is not just one giant thermometer out there, offering up readings that paint a full picture of planetary temps. As we discussed in Chapter 2, it requires a vast network of weather stations around the globe, satellites sending readings back from above, and thermometers on buoys and ships gathering data over the oceans.

What's more, all those measurements coming in from all those instruments aren't perfect. Imagine you placed four thermometers on your kitchen table, evenly spaced apart, and

An ocean-based climate station, maintained by NOAA.

Credit: NOAA News Dec 2010

recorded the temperature from each several times a day. But one of the thermometers sat right underneath a window, meaning sunlight streamed over it for almost all of its readings, while the other thermometers saw sunshine for only part of the day, if at all. Clearly, your four instruments would offer differing readings for the temperature inside your kitchen—not an accurate representation of the situation.

To fix that problem, you could adjust the readings. With careful work, you could figure out a specific amount that the sunny thermometer might have to be adjusted downward in order to let it give an actual, accurate reading of the ambient temperature. If the adjustment was done well, the sunny thermometer would agree with the other three thermometers, since they are all measuring temperatures inside the same small room.

On a global scale, scientists engage in a version of this process known as homogenization. The various weather stations and instrumentation often require adjustment in order to—and this is important—provide accurate readings, not to manipulate the readings to get a desired result. Also, scientists have been taking these measurements for a very long time; the primary record goes back to 1880, with some specific weather stations dating back even farther than that. The physical, logistical situations at those stations can, of course, change over time, again making adjustment necessary. Go back to your four kitchen thermometers: after spending a week recording temperatures, let's say your aunt gives you a lamp for Christmas, which you set on the corner of the table to improve the room's lighting. That lamp is now almost directly over another thermometer, clearly raising its readings; again, an adjustment is obviously warranted to try to get the most accurate measurement.

In the case of weather stations, sometimes buildings go up next door, which could change shadows or even wind patterns around the instrumentation. Scientists again must adjust the readings to correct for such factors.

Back to Homewood and Booker, and the biggest science scandal ever: Homewood reported that he had looked at the raw temperature data for a few weather stations in Paraguay.

He had found that the raw data showed a cooling trend over time, but the adjusted, official data show a warming trend. In other words, those meddling scientists had somehow created warming out of nothing. Shocking! Homewood later went back and found other stations where the same thing had happened: a cooling or flat temperature trend turned into a warming trend, which to him and Booker showed a clear attempt to manipulate scientific data to fit the desired narrative.

Booker really *did* find a cooling trend that turned into a warming trend. Are we sure that's not in the *least* bit scandalous?

Yes. We are sure.

The process of homogenization of temperature data has been well studied, and both the raw and adjusted data are publicly available for others to examine. And others have indeed examined the data. For example, a nonprofit group called Berkeley Earth has found that issues with temperature data "did not unduly bias the record."[5] This assessment may carry more weight than it would coming from another source because Berkeley Earth was founded specifically as a skeptical organization by scientists who had concerns about the consensus surrounding global warming. So they did their own work with the available data and came to essentially the same conclusions as NOAA, NASA, and other major climate science organizations. Richard Muller, the primary scientist involved at Berkeley Earth, even wrote a *New York Times* op-ed entitled "The Conversion of a Climate-Change Skeptic."[6]

Others have also analyzed the homogenization process of temperature data. One study published in the *Journal of Geophysical Research* in 2012 concluded that, if there is any error in the data, it actually has served to *under*estimate the warming trend.[7] In other words, the temperature adjustments have been too conservative, and the real temperature may be even higher than we think.

An earlier paper in the same journal specifically looked for any indication that the weather stations in the United States were inflating temperature trends. Their conclusion, again, was that no such inflation was occurring and that temperature data were robust and accurate.[8]

And here's one more study that shows clearly why Homewood's "analysis" is flawed. In 2011, researchers from NOAA published a paper analyzing the entire Global Historical Climatology Network's data set, which, as the name suggests, is the temperature record for the entire globe. They examined 7,279 total stations that recorded temperature data for some period dating all the way back to 1801, and found that at least one "bias correction" was applied to 3,297 of them.[9] Moreover, those thousands of adjustments to temperature data were not part of a sinister plot to raise the temperatures. The study showed that those thousands of adjustments happened in the positive and negative directions at about an equal rate—hardly the mark of villains creating a warming trend out of nothing.

Of course, all of this evidence doesn't matter when you have a viral article like Booker's, emblazoned with fiery rhetoric, as your source. Congressman Palmer was simply repeating what's available online, and not just at obscure sites but in a major newspaper; it was a subtle bit of wordplay on Palmer's part to rename the blog posts a "report," to make them sound all the more official.

Even a conscientious, curious listener to that radio show may have had trouble getting to the truth; a particularly insidious part of the BLAME THE BLOGGER is that the original mistakes, misconceptions, or outright lies are likely easily accessible with the simplest Google search. Looking online for something like "manipulated temperature records" or "falsified climate data" takes a reader straight to those inaccurate stories, giving them a further air of credibility.[10]

The Internet lets anyone publish anything at all without verification, and once that "report" gets out there, politicians like Gary Palmer can parrot a bogus talking point even after it's been debunked. After all, the Internet still has his back.

GLOBAL WARMING IS A PRIME candidate for the BLAME THE BLOGGER, as there is simply a monumental pile of, well, crap written about it online. To a politician eager to maintain his or her skeptic cred, the Internet is an absolute gold mine of believable-sounding nonsense about climate change.

Here's another example from former Pennsylvania senator, 2012 and 2016 presidential candidate, and prominent TOAD Rick Santorum. In August 2015, he appeared on Bill Maher's show on HBO and confirmed that he did not actually think climate change is a serious problem:

And I'm not alone. The most recent survey of climate scientists said about 57 percent don't agree with the idea that 95 percent of the change in the climate is being caused by CO_2. . . . There was a survey done of 1,800 scientists, and 57 percent said they don't buy off on the idea that CO_2 is the knob that's turning the climate. There's hundreds of reasons the climate's changing.[11]

Santorum was aiming to poke a hole in a statistic you might have heard, regarding the scientific consensus on climate change. The commonly cited number is 97 percent: only three out of every hundred scientists in the field, the idea goes, do *not* agree that climate change is happening and that humans are primarily responsible. You can find this number on the lips of any number of activists, politicians, even NASA's website.[12] So Santorum's claim that *more than half* of scientists are actually *not* on board with mainstream climate science would be a truly dramatic departure.

Far *too* dramatic, as it turns out. Santorum's statistics sounded unlikely enough that Bill Maher had an appropriately skeptical response: "Rick, I don't know what ass you're pulling that out of." Well, it turns out the "ass" in question is—you guessed it— some random site on the Internet!

The site in question is called *Fabius Maximus*, a blog written by a collection of retired military personnel, finance types, and one or two anonymous writers. These are not climate scientists. In fact, they don't seem to be scientists or statisticians of any kind. Why would a nationally visible politician quote them? Well, when you can get someone else to take serious liberties with data and twist findings into knots for you, the real question is, why not?

In July 2015, the *Fabius Maximus* bloggers posted about a survey conducted by the Netherlands Environmental Assessment Agency.[13] According to *Fabius Maximus*, the survey purportedly showed that a minority of climate scientists agreed with the primary finding of the Intergovernmental Panel on Climate Change (IPCC). Or, as their grabby headline put it, "The 97% Consensus of Climate Scientists Is Only 47%."[14]

Grabby, yes. Realistic? Not even close. The key factor that allowed the bloggers to distort this questionnaire's findings is that the survey (which included 1,868 responses from climate scientists of varying expertise and experience) had some room for nuance—it did not involve simple yes-or-no questions. For example, here is question 1a, about the portion of observed global warming that can be attributed to human-caused greenhouse gas emissions:

1a Attribution

What fraction of global warming since the mid-20th century can be attributed to human induced increases in atmospheric greenhouse gas (GHG) concentrations?

- More than 100% (i.e. GHG warming has been partly offset by aerosol cooling)
- Between 76% and 100%
- Between 51% and 76%
- Between 26% and 50%
- Between 0 and 25%
- Less than 0% (i.e. anthropogenic GHG emissions have caused cooling)
- There has been no warming
- Unknown due to lack of knowledge
- I do not know
- Other (please specify)

Credit: Netherlands Environmental Assessment Agency

The second question, which was dependent on the response to the first, asked respondents to ascribe a confidence level to their answer. The *FM* bloggers' primary bit of statistical mischief involved two of the possible responses to question 1a: "I do not know" and "Unknown." The bloggers assumed that those answers—accounting for 9.9 and 8.8 percent of those who responded to the question—meant the respondents did not agree with the basic premise that humans are primarily responsible for climate change. The actual authors of the study, though, disagreed. They pointed out that selecting those less certain responses does not necessarily mean that respondents don't agree with the premise, but instead could indicate that respondents believe pinpointing the exact amount of warming caused by humans could be difficult.

That disagreement, as it turns out, completely changes the conclusions of the survey. The *FM* bloggers took only those respondents whose answer to question 1a was that the human contribution exceeded 50 percent *and* who then said it was either "virtually certain" or "extremely likely" that the 50 percent mark was true, and divided that number (which turned out to be 797 individuals) into the full 1,868-scientist cohort. The result: 43 percent agree with the consensus, and 57 disagree. Voilà, a Rick Santorum sound bite.

There was actually another problem with the *Fabius Maximus* analysis: by allowing only "virtually certain" or "extremely likely" in question 2, they misstated the actual current state of consensus. See, the Netherlands survey was conducted in 2012, *before* the release of the latest IPCC report. The previous report, from 2007, maintained the official conclusion that it was "very likely" that human-caused emissions had caused most of the observed warming; it was only with the 2013 version that "extremely likely" became the consensus finding. These terms carry specific meanings: "very likely" means greater than a 90 percent probability; "extremely likely" means more than 95 percent; and "virtually certain" means 99 percent or above. If the *FM* writers wanted to be accurate, they should have included the survey respondents who said it was "very likely" as well; that would have raised their conclusion from 43 percent all the way to 67 percent.

The original authors of the survey arrived at very different—and higher—numbers by being more careful and precise. They excluded those "I don't know" and "Unknown" answers entirely and then divided up the respondents according to how many papers they had published on climate science—a crude but reasonable method for determining how much a person knows about the field. They found a range of consensus from 79 to 97 percent; the lower number was only among those who had published zero to three papers, while the highest percentage was among those intimately involved with the IPCC process. One might call the latter group "experts."

In total, the percentage agreeing with the consensus was 84 percent—a far cry from the 57 percent *dis*agreement Santorum claimed. Yet the deviousness of the BLAME THE BLOGGER was clear from Maher's response: even someone who is aware of the science, and wholly skeptical of the denier claims, couldn't immediately rebut Santorum's point. Average viewers of the show likely wouldn't know where to look to see whether Santorum was correct, and again, a Google search for related terms might lead them to the incorrect source. Santorum could crouch behind the Internet, comfortable that few would do the math to peel back the curtain on his hiding place.

———

SOMETIMES THE "BLOGGER" in question is the author of a 1975 magazine article. Yes, this is a bit of a deviation from the narrative that has the Internet's level playing field as a culprit, but the "global cooling" myth persists less because of the specific articles from decades ago and more from the recurring citation of them by writers online today.

The common refrain is that in the 1970s, scientists warned of a coming apocalypse: the world was cooling down thanks to human-caused emissions—so much so that we could be headed for a new ice age that would kill crops and render large parts of the world uninhabitable. Obviously, that dire prediction didn't come to pass, and the scientists all shifted their viewpoint from cooling to warming. Why should we trust them now if they just love to shout that the sky is falling, picking a new bogeyman every few decades?

The answer is that very few actually *were* concerned about global cooling, and though the science was still in its infancy, even in the 1970s most were more concerned that a warming trend was on the horizon than the opposite. But that hasn't stopped politicians from squawking about this. Senate majority leader Mitch McConnell has mentioned it, as we learned in the Introduction, among many others. Here's Ted Cruz with another version:

> I read this morning a *Newsweek* article from the 1970s talking about global cooling. And it said the science is clear, it is overwhelming, we are in a major cooling period

and it's going to cause enormous problems worldwide. . . .
Now, the data proved to be not backing up that theory.
So then all the advocates of global cooling suddenly
shifted to global warming.[15]

There was indeed a *Newsweek* article that said essentially those
things—a single-page piece published on April 28, 1975.[16] It
was, in fact, quite ominous in tone, but the story's author himself, a journalist named Peter Gwynne, has since thoroughly
distanced himself from it. Writing for Inside Science in 2014,
Gwynne said:

> Here I must admit *mea culpa*. In retrospect, I was over-
> enthusiastic in parts of my *Newsweek* article. Thus, I
> suggested a connection between the purported global
> cooling and increases in tornado activity that was unjus-
> tified by climate science. I also predicted a forthcoming
> impact of global cooling on the world's food production
> that had scant research to back it.
>
> The messages for science writers are to ask questions
> beyond the obvious and to seek out what the science
> doesn't imply as well as what it does. If I had applied
> those lessons back in 1975, I might not now be in the
> embarrassing position of being a cat's paw for denial of
> climate change.[17]

Gwynne added that "those that reject [current] climate science
ignore the fact that, like other fields, climatology has evolved
since 1975." In short, the article was wrong, both in content and

in tone. (For completeness, *Time* also published a global cooling article, in 1974,[18] that occasionally is brought up in a similar fashion. As we'll see, neither article was properly representing what scientists actually thought.)

The fundamental question here is not whether popular magazine articles were overblown, but whether the scientists studying these topics believed what Cruz says they believed. And interestingly enough, some actual research has looked into this question. In a literature review published in the *Bulletin of the American Meteorological Society* in 2008, researchers assessed how many peer-reviewed papers published between 1965 and 1979 seemed to favor a global cooling scenario, a global warming scenario, or "neutral" temperature trends.[19] The results were clear.

Throughout that period, fewer than ten peer-reviewed papers were published with cooling as the focus. Meanwhile, twenty "neutral" papers appeared, and forty-four papers indicated a warming trend; by the end of the 1970s the slant toward papers on warming seemed more pronounced, with no cooling papers at all appearing in 1978 or 1979. Does this sound like "all the advocates of global cooling" were a majority, who later just jumped ship and switched to Team Warming?

In reality, the scientific community was already well aware of the impending warming problem. A 1979 report from the National Research Council, which is part of the National Academies, already issued fairly dire warnings regarding the warming world: "A wait-and-see policy may mean waiting until it is too late."[20] Humanity may want a do-over on that one, unfortunately.

Just as with the more modern examples of the BLAME THE BLOGGER, McConnell, Cruz, and others who beat this particular horse rely on the fact that few are going to check their work. Online, it is incredibly easy to find the "global cooling" scares from *Time* and *Newsweek*, as well as comments from countless TOADS blogging about them in recent years, and far more difficult to find literature reviews or actual studies from the 1970s poking holes in the idea that an enormous consensus had formed. Climate science has, of course, progressed since 1975, but arguing that a couple of magazine articles somehow undercut the entire scientific enterprise is a bit beyond the pale.

IT IS NOT JUST CLIMATE CHANGE—and it is not just blog posts or news articles—that can yield BLAME THE BLOGGER fodder for opportunistic politicians. In the summer of 2015, the Internet exploded with news of undercover videos purporting to reveal some shocking practices taking place at Planned Parenthood clinics. Politicians on the right seized on these videos with vigor, repeating some truly incendiary rhetoric and making claims that the facts simply couldn't support.

The videos, made by a group run by anti-abortion activist David Daleiden, called the Center for Medical Progress (CMP), supposedly unveiled the shady world of "selling baby parts"— the concept that abortion providers were taking aborted fetal tissue and profiting off its sale to researchers, sometimes without consent of the mothers. This was a drastic and utterly morbid twisting of the truth, and one that arguably began a backlash against Planned Parenthood that resulted in violence. The lack of truth in the videos, however, did not stop politicians such as

Carly Fiorina, Rand Paul, and Rick Perry from citing them as fact. Again: it's on the Internet; therefore it's true.

There were many specific allegations and claims made regarding the Planned Parenthood videos, but the most fundamental and oft repeated was that the clinics were making a profit off the sale of aborted fetuses. Here is Rick Perry in July 2014:

> The video showing a Planned Parenthood employee selling the body parts of aborted children is a disturbing reminder of the organization's penchant for profiting off the tragedy of a destroyed human life.[21]

Carly Fiorina made a similar claim:

> This latest news is tragic and outrageous. This isn't about "choice." It's about profiting on the death of the unborn while telling women it's about empowerment.[22]

Rand Paul chimed in as well, tweeting out supposedly incriminating quotes from "a video showing [Planned Parenthood]'s top doctor describing how she performs late-term abortions to sell body parts for profit!"[23]

And so on; other politicians made similar statements regarding sale and profit. The supposed evidence for such claims was in the highly edited videos of Planned Parenthood executives and doctors having lunch with a fake tissue procurement organization that Daleiden had created (using a fake name). The videos, running about ten minutes long, based on hours-long raw footage, used dramatic on-screen text and black-and-white replays and generally painted a picture of a singularly evil organization hell-bent on creating a black-market trade in dead babies. It was convincing propaganda.

Several lines in the videos inflamed the politicians on this issue. The first and most widely disseminated video features a Planned Parenthood doctor named Deborah Nucatola. In what is indeed a cavalier and somewhat careless tone, she is shown discussing the dollar amounts that a clinic should be reimbursed per fetal tissue specimen; specifically, she cites a range of $30–$100.

On its own, this statement makes for a compelling case. Perry, Paul, and Fiorina didn't bother discussing the details of those dollar amounts, instead focusing on the concept of sale or profit. But the edited videos declined to show a whole host of other lines from Nucatola decrying the entire idea of profit, instead insisting that clinics are only looking to recover costs.

Here's the important, fundamental point about all of this: fetal tissue donation from abortions is *legal*, and financial remuneration to cover costs—but *not* to profit—is also perfectly allowed under federal law. The reason it is legal is that fetal tissue has historically been a hugely useful source of scientific discovery. Fetal cells are in some ways uniquely suited for certain research pursuits. Here's how the American Society for Cell Biology describes their utility:

> Fetal cells hold unique promise for biomedical research due to their ability to rapidly divide, grow, and adapt to new environments. This makes fetal tissue research relevant to a wide variety of diseases and medical conditions.[24]

Fetal tissue has been used to develop vaccines against measles and rubella, and even to help grow the polio vaccine. The 1954 Nobel Prize in Physiology or Medicine was awarded to three scientists—John Franklin Enders, Thomas Huckle Weller, and Frederick Chapman Robbins—who discovered that the poliovirus could grow in fetal tissue cells.[25] Their work showed that large quantities of the polio vaccine could be grown quickly—an important step in eradication of the disease in the United States.[26]

Many other examples illustrate the usefulness of this type of tissue. The legality of fetal tissue donation was enshrined in a 1993 law called the National Institutes of Health Revitalization Act.[27] Among its various provisions was a section on fetal tissue donation, making it legal as long as certain requirements are met. The law specifies that "valuable consideration" for the tissue donation is not allowed, and goes on to make exceptions to that term:

> The term "valuable consideration" does not include reasonable payments associated with the transportation, implantation, processing, preservation, quality control, or storage of human fetal tissue.

Okay, but Dr. Nucatola gives some specific numbers in the CMP video; do they represent "reasonable payments" or not? As it turns out, the answer is a fairly unequivocal yes.

A series of experts chimed in to FactCheck.org about what a reasonable amount of money might be in this scenario, and all agreed: the prices quoted in the video could in no way represent profit.[28] Here's Carolyn Compton, a former director of biorepositories and biospecimen research at the National Cancer Institute, and the chief medical and science officer at Arizona State University's National Biomarker Development Alliance:

> "Profit" is out of the question, in my mind. I would say that whoever opined about "profit" knows very little about the effort and expense involved in providing human biospecimens for research purposes.

And here's Sherilyn Sawyer, the director of Harvard University's "biorepository":

> There's no way there's a profit at that price. . . . In reality, $30–100 probably constitutes a loss for [Planned Parenthood]. The costs associated with collection, processing, storage, and inventory and records management for specimens are very high. Most hospitals will provide tissue blocks from surgical procedures (ones no longer needed for clinical purposes, and without identity) for research, and cost recover for their time and effort in the range of $100–500 per case/block. In the realm of tissues for research $30–100 is completely reasonable and normal fee.

Objective research into the topic of fetal tissue procurement is limited, but there is evidence that those various experts are correct in their assessments. In a report released by the US General Accounting Office (later renamed the Government Accountability Office) in 2000, the GAO determined that, generally speaking, clinics did not charge researchers looking to use fetal tissue anything at all. When they did charge, the price ranged from a low of $2 to a high of $75, and it averaged $22.[29]

Add all this up—along with the fact that an extremely limited number of Planned Parenthood clinics in only two or three states have ever even been involved in fetal tissue donation—and it becomes clear that the accusations presented by the Center for Medical Progress and repeated by the GOP politicians were utterly hollow. But that didn't matter to those politicians, especially given how contentious the issue of abortion is in this country.

They continued their fierce attacks for months following the release of CMP's videos, always falling back on the idea that there was a *source* for their claims. "You can go watch the videos yourself, right there on the Internet! How can you deny what's in them?" And it wasn't just empty rhetoric: many in the GOP actually threatened to shut down the government if federal funds continued to flow to Planned Parenthood, ignoring the fact that no laws had been broken (multiple state investigations found no evidence of wrongdoing of any kind[30]), and Planned Parenthood is actually barred from using its federal funding specifically for abortions.

During one GOP debate in September 2016, former HP CEO Carly Fiorina took this particular BLAME THE BLOGGER to its extreme with a fiery description of what sounds like a truly horrific scene:

As regards [to] Planned Parenthood, anyone who has watched this videotape, I dare Hillary Clinton, Barack

Obama to watch these tapes. Watch a fully formed fetus on the table, its heart beating, its legs kicking while someone says we have to keep it alive to harvest its brain![31]

By many accounts, Fiorina "won" the debate with this sort of incendiary rhetoric. The only problem is, that scene she described doesn't exist, and never did.

One of the various CMP videos contains a short clip of what appears to be a not-yet-fully-formed fetus on a table, with "its legs kicking." That same video contains a *description* of the scene Fiorina mentions, regarding harvesting a brain, by a former employee of a tissue procurement company. There is no way to verify the interviewee's story, and the company she worked for vehemently denies any wrongdoing.[32]

The short clip of a fetus, it turns out, did not even come from CMP's undercover footage. It came from a pro-life group called the Grantham Collection, and the Center for Bio-Ethical Reform. Those groups would not tell media where exactly the footage was shot, though they claimed that the writhing fetus was, in fact, the result of an abortion. Obstetrics experts, though, chimed in later, asserting that there was almost no chance the video showed an aborted fetus, but instead showed a miscarriage.[33] Regardless, the scene as Fiorina described it—with evil, supposedly cackling demons insisting the fetus be kept alive to "harvest its brain"—simply never occurred. Daleiden and the CMP, of course, backed her in full—the blamed blogger, in this case, was on board with the politician.

A terrible postscript to this story took place a couple of months later: although a hard cause and effect can't be determined, some felt that the inflammatory reactions to the CMP videos led to an act of domestic terrorism. Robert Dear, a possibly mentally ill man, killed several people at a Colorado Springs Planned Parenthood clinic and reportedly said "no more baby parts" upon being captured by police.[34] At a court hearing, he shouted: "I am a warrior for the babies."[35]

The resultant violence is rare, but letting politicians spout whatever nonsense the Internet offers without challenge is indeed a slippery slope. The BLAME THE BLOGGER is in some ways a free pass for politicians to lie; if they are just quoting someone else's claim, can you really blame them? We *should* blame them; politicians have a public influence that the average blogger does not wield, and they should be held to a higher standard when it comes to science. Calling out their repetitions of silly, ridiculous, or downright dangerous Internet nonsense can have wide-ranging impacts: stopping the spread of abortion-related misinformation could help reduce the possibility of extremist violence, for example, and building trust in climate science could help push for action, quite literally helping to save the world.

CHAPTER 6

The Ridicule
and Dismiss

WHICH SCARES YOU MORE—A BEHEADING OR A SUNBURN? I'm sure you will debate the answer to this carefully—*well, a beheading happens really fast if done properly, just blink and it's all over, while a really bad sunburn is extremely uncomfortable and could actually substantially raise my risk of eventually developing melanoma, and though advances in treatment for late-stage melanoma have certainly been coming fast and furious in recent years, it is still an incurable disease with extremely poor prognosis and seems a fairly excruciating way to die*—but it's okay, you can give the easy answer. Aloe soothes only one of those two.

In early 2015, former Arkansas governor Mike Huckabee began rolling out a line in speeches designed to do two things at once: criticize President Obama's treatment of radical groups in the Middle East, and ridicule the president's recent focus on combating climate change. Here's how Huckabee put it at an event in Iowa, in January, a few months before he declared his candidacy for the presidency:

When [Obama] said: "The greatest threat this nation faces . . . is climate change." Not to diminish anything about the climate at all, but Mr. President, I believe that most of us would think that a beheading is a far greater threat to an American than a sunburn![1]

He used the line more than once. Huckabee was banking on the fact that his target audience was deeply concerned about the spread of ISIS, Al Qaeda, and other threats halfway around the world, but essentially not concerned at all about climate change. He tried to enhance that perceived divide even further by describing the scientific issue in such ridiculous fashion as to make it nearly impossible for the relatively underinformed to disagree. *A sunburn?! The president is being ridiculous! How can ISIS be less of a threat than climate change? How can climate change be considered a threat at all?*

That's the RIDICULE AND DISMISS, the technique of making a complicated scientific topic sound so silly that the audience members can only shake their heads and laugh. A political point is made at the expense of people actually understanding the scientific topic in question.

Huckabee managed to belittle something *and* completely mistake the science at the same time. A sunburn, though of course something to be avoided, certainly doesn't sound all that scary. But a sunburn has essentially *nothing* to do with climate change.

We get a sunburn when ultraviolet radiation from the sun—or a tanning bed! (Don't use tanning beds. Seriously.)—causes damage to DNA in our skin, provoking a molecular reaction and resulting in that familiar painful burn. But clearly, you get more sunburns when it is *sunnier* outside, not *hotter*. Just think of skiers and snowboarders who put on sunscreen on a sunny winter day; they know that even in those 20-degree temperatures, the bright sun can fry your face just as quickly.

The president, in his State of the Union address in January 2015, and again at other times, has indeed called climate change one of the greatest threats facing the world, and facing future generations.[2] He has not, however, reminded everyone to cover up at the beach or on the slopes and wear sunscreen. These issues are totally, completely, unrelated.

Okay, not *totally* unrelated. Here's the only possible way Huckabee could have defended this bit of misdirection. Remember the ozone hole? Different environmental issue entirely, right? Well, there is a connection: chlorofluorocarbons, hydrochlorofluorocarbons, and the related gases primarily responsible for causing the hole in the ozone layer, which was big news in the late 1980s and early 1990s, are *also* greenhouse gases, meaning that just like carbon dioxide, they trap heat when in the atmosphere. In fact, they trap heat *thousands of times* more effectively than carbon dioxide does.[3] We emit a far smaller volume of these gases than of carbon dioxide or methane (and the world has made substantial progress toward phasing out their use entirely), so this potency is somewhat attenuated by the smaller amounts, but they are still important warming agents.

So, let's connect the dots for Governor Huckabee: Gases that help cause global warming also cause a hole in the ozone layer. A hole in the ozone layer, if above your head, would let more UV rays hit you and increase your risk of . . . you guessed it, sunburn. If that sounds a bit *thin* . . . well, yes. It is.

I think it's safe to say that Governor Huckabee was not trying to connect the continued use of fluorinated gases to climate change by mentioning sunburns. Instead, he was mistaking one thing (the world getting warmer) for another thing that sort of sounds like it might be connected (sunny days and sunburn). Reducing climate change and all its attendant, potentially catastrophic effects to a sunburn makes it sound silly enough, even before comparing it to the very scary idea of ISIS beheading you and broadcasting the video on the Internet.

Contrary to Huckabee's RIDICULE AND DISMISS, however, climate change does indeed represent a significant threat to countries and individuals around the world. And since he brought it into the realm of the military by comparing his not-at-all-relevant sunburn to a beheading by militant groups in the Middle East, let's go straight to a very relevant source to see why he's wrong: the Pentagon.

Though much of the US government has dragged its heels on addressing the issue of climate change, the military has, in fact, been calling for action on the issue for years. Without any elections to run, military leaders seem free to look at the science without bias, and they see a dire situation.

The Pentagon released a "Climate Change Adaptation Roadmap" in October 2014,[4] calling global warming a "threat multiplier." This means it could exacerbate existing problems; if an issue we face is a 3 out of 10 on the "Should We Be Worried?" scale, a threat multiplier might turn it up to 6 or 7.

How exactly would that happen? Here's what the Pentagon wrote in that report:

Rising global temperatures, changing precipitation patterns, climbing sea levels, and more extreme weather events will intensify the challenges of global instability, hunger, poverty, and conflict. They will likely lead to food and water shortages, pandemic disease, disputes over refugees and resources, and destruction by natural disasters in regions across the globe.

Now *that* sounds terrifying. Basically, every global and geopolitical issue you can think of will potentially be made worse by a warming planet. Almost as if it were a direct rebuke to Huckabee's rhetoric, the Pentagon report even specifically cited terrorism as a threat that climate change could "exacerbate." Here's the report's explanation of how that might happen:

The impacts of climate change may cause instability in other countries by impairing access to food and water, damaging infrastructure, spreading disease, uprooting and displacing large numbers of people, compelling mass migration, interrupting commercial activity, or restricting electricity availability. These developments could undermine already-fragile governments that are unable to respond effectively or challenge currently-stable governments, as well as increasing competition and tension between countries vying for limited resources. These gaps in governance can create an avenue for extremist ideologies and conditions that foster terrorism.

Once again, terrifying. Some research has even connected climate change to very specific geopolitical crises, such as the civil war in Syria—the exact place where ISIS and its penchant for beheadings got its start. A study published in the *Proceed-*

ings of the National Academy of Sciences in early 2015 found that "anthropogenic forcing"—meaning human influence—made the drought that Syria experienced from 2007 to 2010 as much as three times more likely than by natural variability, and that drought contributed to the subsequent unrest.[5] "We conclude that human influences on the climate system are implicated in the current Syrian conflict," the authors wrote, meaning that yes, climate change arguably played a part in those beheadings broadcast on the Internet.

So, to recap: Mike Huckabee completely misstated what climate change actually *is*, comparing it to a thing you get from standing out in the sun for too long. And he belittled its importance by comparing it to terrorism—which American military leaders have specifically said could be made worse by climate change.

He got some cheers, of course, and made his point: the president is weak on the *real* issues (the radicals that commit atrocities and broadcast them online), and he is strong on this *fake* issue (this silly climate thing that won't actually affect us in any meaningful way). In a sense, Huckabee gave his audience the easy way out; he gave them permission to ignore the complexities of climate science, by making it all sound ridiculous.

NOT EVERY VERSION of the RIDICULE AND DISMISS involves terrifying ways to die; politicians can also belittle and undermine life-and-death science without playing to your fears. Let's switch from beheadings to fruit flies.

For a few years, former Oklahoma senator Tom Coburn released what he called the "WasteBook"—a list of some of what he and his staff considered the most egregious bits of "pork" on the government ledger. Coburn, consistent with the standard GOP platform, wanted to rein in government spending, and he used the "WasteBook" as a way to highlight how much is squandered.

The table of contents of one of these books alone makes for a great read. Here are some of the science-ish selections from the 2012 version, ostensibly descriptions of things the government actually spends money on: "Out-of-this-world Martian food tasting. . . . When robot squirrels attack. . . . How to build a farm in a galaxy far, far away . . . Crazy for cupcakes . . ."[6] And so on. Each "WasteBook" generally featured a hundred entries.

Entry number 70 in that 2012 version is titled "Fruit fly beauty is fleeting." This bit of near-Pythonian nonsense found its way into a speech given by Kentucky senator Rand Paul in the early days of 2015. Though Coburn's book version had time to actually explain what it was talking about in some detail, here's how Paul described it, after mistakenly saying the budget at the NIH, the National Institutes of Health, had been rising for years (it hadn't):

> But you know what they did discover? They spent a million dollars trying to determine whether male fruit flies like younger female fruit flies. I think we could have polled the audience and saved a million bucks![7]

Just like Huckabee and his sunburn, Paul took a complicated topic—we'll get to it—and made it sound outlandish. Why would the government spend *any* money, let alone a *million* dollars, to investigate fruit fly sexual proclivities?

If that's all the information you have, it sure does sound absurd—and Paul got a good laugh from his audience about it. But Paul was engaging in a double-layered RIDICULE AND DISMISS: he made light of a particular study that is far more important than he let on, and more generally, he made it seem as though studying anything at all with fruit flies would be a waste of money. This is about as far from the truth as you can get.

First of all, the specific research Paul was talking about was part of an ongoing series of studies in the lab of a professor named Scott Pletcher, now at the University of Michigan. Pletcher's lab examines sensory perception and olfaction, the aging process, and how these things relate to sexual and social activity.[8] To study these issues, the lab uses fruit flies as a "model organism." A model organism is, basically, a substitute for humans; many of the things we want to know about ourselves would be impossible to actually study in people, so we use smaller, less complicated animals instead.

This approach isn't perfect. Just because, say, a cancer drug works in mice does not necessarily mean it will work in humans. But before we test that drug on real people with cancer, who are desperate for anything to extend their lives, it's good to know that at least in some other creatures with important similarities in terms of genetics and anatomy, it does seem to work.

In the case of the fruit fly, we're not testing drugs, generally. Fruit fly research is more fundamental, aiming to understand cells, parts of the brain, connections between those parts, and various important proteins—some of the most basic biological functions of living organisms.

Pletcher's lab has done some important work, reflected in

particular by one paper the group published in the prestigious journal *Science* in 2013. The research showed that exposing male flies to female pheromones without giving them the opportunity to mate—imagine watching the object of your desire walking past you over and over, on the other side of a one-way mirror—had decreased life spans.[9] Pletcher said in a press release at the time that, in other words, sexual reward "specifically promoted healthy aging."[10] This is not without relevance to people, obviously.

The specific finding that Paul sarcastically attacked was related. It focused on the decline of those pheromones in female flies over time, which led to the male flies trying to mate with younger females.[11] Though again, this may sound silly, the point, once more, is to learn something about ourselves, or other animals, or the world in general from these little critters. That paper's abstract concludes that production of those pheromones, or similar compounds, "may be an honest indicator of animal health and fertility." Does that sound like as much of a waste?

Paul's point is that this particular million dollars is representative of larger wasteful tendencies. He suggested that the budget for the NIH, which, along with the National Science Foundation, accounts for the bulk of US dollars going into basic science research, is too high. As we saw in Chapter 3 (the BUTTER-UP AND UNDERCUT), the NIH actually has an annual budget that has stagnated at about $30 billion—less than 1 percent of the total federal budget—since 2003 (and declined when inflation is considered).

And though the NIH's budget stagnated under George W. Bush, there is increasing understanding that supporting many types of scientific research does not have to be a partisan issue. In fact, former Republican Speaker of the House Newt Gingrich published an op-ed in the *New York Times* in 2015, calling for a doubling of the NIH budget in order to help prevent disease and save far more money later on. "We are in a time of unimaginable scientific and technological progress," he wrote. "By funding basic medical research, Congress can transform our fiscal health, and our personal health, too."[12]

If politicians resort to sarcastically deriding the basic research that Gingrich is talking about, public support will never shift in favor of science.

But making fun of one study was just the first, surface level of Rand Paul's derision. By belittling Pletcher's work, he managed to belittle fruit fly research in general, ignoring more than a century of crucial scientific research and advancement. As any

The fruit fly, *Drosophila melanogaster.*

biologist, neuroscientist, pretty much any other type of scientist, and of course large chunks of the general population could probably tell you, the fruit fly is *super* important.

Scientists began using the fruit fly—Latin name *Drosophila melanogaster*—in the first decade of the twentieth century. Here's fly expert Hugo Bellen, of Baylor College of Medicine, with a good rundown of some of the fruit fly's accomplishments:

> It has been pioneering research. . . . [*Drosophila*] has led to the discovery of genes that cause cancer, genes that [affect] metabolism, genes that cause developmental defects, genes that play a critical role in neurodegeneration. It has been a discovery tool for many, many different pathways, proteins, diseases.[13]

Impressive! Bellen also explained some of the reasons why such a tiny creature, so foreign-looking when compared to humans, could help us understand so many new things about ourselves. Among the most important points is what's known as the "conservation" of the fly's genetics in humans. Of the fruit fly's approximately fourteen thousand genes,[14] about eight thousand are "conserved" in humans. Essentially, we have a whole lot of the same genes that a fruit fly has. Therefore, studying how these genes operate in flies can give us a pretty good idea of how they also operate in humans. For example, researchers can create flies in which a specific bit of genetic code is "knocked out," so that a certain protein is not produced (or is overproduced); the resulting "phenotype"—basically, how the fly without that gene looks and behaves—can give us information about what that gene does.

Combine these ideas with the particulars of the flies themselves—they breed, are born, and die very quickly—and you've
got essentially a perfect model organism for understanding
basic neuroscience.

Even this wonderfully useful organism, however, can sound
absolutely ridiculous if described in certain ways. Paul made
that clear with his remarks. But ridiculing fly research is actually really easy; we can all do it! How about this: *Scientists spent
years counting up how many fruit flies have red eyes, and how many
have white eyes.*

Sounds like an absurd waste of time! But that's a fairly accurate description of a series of experiments led by pioneering fly
researcher Thomas Hunt Morgan. By counting the numbers of
flies that had white eyes rather than red, Morgan was able to
prove that the genes are passed down from generation to generation on chromosomes, organized like beads on a string.[15]
He essentially discovered exactly how you get traits from your
parents. For his efforts, Morgan won the Nobel Prize in Physiology or Medicine in 1933.[16]

Let's keep going: *A scientist irradiated male fruit flies and then
had them mate to see what happened.* This is not a setup for a sci-fi
movie. (Though it probably should be.)

This silly-sounding experiment may have even more practical import than Morgan's work on heredity. Hermann Müller,
in 1926 and 1927, showed that X-rays can cause mutations to our
genes. The flies were subjected to doses of radiation and then
mated to female flies; their progeny exhibited a wide range of
mutations, some deadly and some not.

This is, essentially, why you wear a lead apron when you
get an X-ray. But not only that; Müller's research showed us

that genetic manipulation is possible, since the mutations were passed on to subsequent generations. This finding had far-reaching implications, not just in medicine, but in fields like agriculture as well. Müller also won the Nobel Prize in Physiology or Medicine for his work, in 1946.[17]

Fly work hasn't stopped there. At least two other Nobels have gone to research directly involving fruit flies,[18] and many others have arisen out of the basic science that flies helped illuminate. To this day, *Drosophila melanogaster* remains among the most important model organisms, along with mice, rats, and a few others. They are involved in research into some of our most pressing biological and medical questions, including the neurobiology involved in nervous system diseases like Alzheimer's.

Paul's brand of the RIDICULE AND DISMISS provides a clear example of how science can be misused in the hands of politicians. Though he mentioned only one particular study—and mischaracterized it—Paul (who is, by the way, a doctor, and should know better) managed to dismiss a century's worth of important science with the wave of his hand, just for a cheap laugh. One would not expect the average person listening to a political speech to know about Thomas Hunt Morgan and Hermann Müller and the genetics of *Drosophila melanogaster*— and after hearing lines like that, why would the general public come away thinking that supporting scientific research is a good idea at all?

The details of government spending are, obviously, up for debate. Maybe Paul would prefer that other areas of science receive more support. But even that is a bit shortsighted: Bellen, the fly expert at Baylor, said: "You get 10 times more biology for a dollar invested in flies than you get in mice."[19] That's thanks to the ease with which we can manipulate their genes, and the speed at which they reproduce. If you're belittling something to make a point, it helps to actually *know* about the thing you're belittling.

THIS SORT OF ANTISCIENCE SENTIMENT—or, at best, misunderstanding of scientific method and process—has been around for decades, if not longer. Years before Senator Coburn began his "WasteBook," Democratic Wisconsin senator William Proxmire had a similar project called the Golden Fleece Awards. The premise was essentially the same: give the "award" (handed out semiregularly from 1975 through 1988) to a particularly egregious bit of government waste, and describe the winning work in a way that made it nearly impossible to defend. Many of the targets were scientific in nature, and such was Proxmire's influence that a Golden Fleece sometimes was equivalent to a death knell for your research.

Just as with Coburn and Paul, occasionally Proxmire was way off base with his criticisms. For example, he called out the National Institute on Alcohol Abuse and Alcoholism in 1975 for funding a study on aggression related to alcohol. On its own, such a study sounds reasonable, but this one was conducted in fish and rats—an easy joke to make. One of the researchers, though, pointed out some of the absurdity of attacking science in this way: "I would really enjoy having Proxmire make a proposal to give people alcohol and ask them to fight."[20]

In recent years, some have come up with antidotes to the "WasteBooks" of the world. The Golden Goose Awards (established in 2012) aim to show that even silly-sounding science sometimes yields big returns; if we ridiculed and ignored all of it, some major discoveries would never have happened. For example, a 2014 award went to researchers for a technique involving massaging rat pups. According to the awards group, that technique "led to a momentous change in how premature babies are cared for that has saved lives and billions of dollars in health care costs. Because of this research, thousands of preemies have survived, grown stronger, thrived, and gone on to live healthy lives."[21] Sounds a lot more "Goose" than "Fleece" now, doesn't it?

In some ways, science is an easy thing to ridicule. A lot of scientific research is basic, and simple, and adds up to something relevant and practical only when each layer is added on to many earlier layers. As such, it's an easy target for a politician. Just as we discussed with the OVERSIMPLIFICATION, sound-bite politics can harm the actual practice of science. The tactic of obscuring the importance of a scientific issue, be it climate change or fly research or rat pup massages, through sarcasm and quick quips may garner some applause at a stump speech, but it also actively erodes the public's understanding of and appreciation for science.

....

CHAPTER 7

The Literal Nitpick

TWO PARENTS SIT IN THE KITCHEN, ENJOYING THEIR MORN-ing coffee. A crash is heard from the other end of the house. They look at each other and head toward the noise. They find a broken window near a baseball, and their two children sitting on the other side of the room, supposedly engaged in a book together. The kids very purposefully do not look up.

The father sighs. "Everyone okay?"

The kids nod slightly.

The mother asks, "Did you guys break the window?"

The children look up, triumphant. "No! Of course we didn't break the window."

The parents look at each other, eyebrows raised. "No? Then how did it break?"

"The *baseball* broke the window. Not us."

The kids in this scenario are, of course, correct in literal fact. The baseball *did* break the window, and these burgeoning logicians took advantage of the literal point to attempt their getaway. The parents, though, are unlikely to buy that bit of semantic maneuvering. The baseball didn't throw itself at a window, after all.

As silly as this example may be, politicians use exactly the same technique when it suits them. Perhaps the best recent example involves a contentious issue in American policy: fracking. Here's Senator James Inhofe, in a press release regarding some regulations and legislation surrounding fracking for natural gas:

> Since 1949, my state of Oklahoma has led the way on hydraulic fracturing regulations, and just like the rest of the nation, we have yet to see an instance of ground water contamination.[1]

As we'll see, this statement is essentially equivalent to "the baseball broke the window." It's an example of the LITERAL NITPICK, wherein a politician homes in on the very specific definition of the words involved, or a few particular words among many, in order to stay out of trouble. If called out on the issue, the politician has a ready-made response: "I was literally correct! Check the actual words I said!" This caginess can be hard to see through; if you don't know much about fracking, Inhofe's statement sure sounds airtight. Not a single instance of water contamination? Well, then it must be safe, right? A more reasonable reading would say no—*someone* had to throw that baseball.

First, a bit of background on fracking. The word (sometimes written "fracing") is shorthand for hydraulic fracturing, a technique used to extract oil and gas from deep underground. It was first developed in the 1940s, as Senator

Inhofe mentioned, but it became crucially important thanks to technological advances in the first few years of the twenty-first century. Along with another technique, known as horizontal drilling, fracking helped spawn the massive oil and gas booms in Pennsylvania, North Dakota, and elsewhere around the country.

The idea of fracking is to break apart rocks in order to release the oil and gas hiding in tiny cracks and fissures. First, a well is drilled, much as with conventional oil and gas extraction. Then, drillers send a mix of many thousands of gallons of water, sand, and a combination of chemicals known as fracking fluid down into the well. This combination exerts

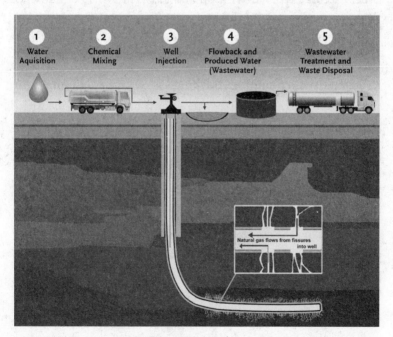

The process of fracking and the associated water cycle.

pressure on the rocks and breaks them apart, allowing the gas and oil to seep out, which is then pulled back up to the surface. By the time Inhofe released his statement, about 90 percent of all new oil and gas wells drilled on public or tribal lands were using fracking.[2]

Inhofe's statement was a response to a regulation, released by the Department of the Interior, specifically relating to wells drilled on public and tribal lands. The rule contained a number of provisions, such as a requirement that companies disclose exactly which chemicals make up their fracking-fluid soup, another requiring the validation of the well's cement integrity, and more detailed information disclosure on the geology and other factors involved with oil and gas drilling. The industry, of course, opposed these rules, and most of the Republican Party was on its side.

The main issue that led to the rule, and the more general controversy surrounding fracking, is water contamination. Since the fracking boom began, many have been concerned that the increase in oil and gas activity could contaminate water supplies, both in rivers and lakes and in groundwater. This could happen in a number of ways: fracking fluids or natural gas itself could somehow escape the well as it is coming up or down from belowground; those same fluids could spill at the surface during the course of fracking-related activities; or, most difficult to assess, the fluids could theoretically seep up from far belowground through cracks in the rocks until they reach the water table. Water contamination, in general, is the primary issue behind Josh Fox's Oscar-nominated documentary *Gasland*, which features some memorable footage, such as tap water catching fire.

Many experts, including those at the Environmental Protection Agency and others at universities and institutes around the country, have been trying to gauge water-related fracking risks since the boom began. This has proved exceptionally difficult to do, in part because the oil industry has made it so. For many years, oil and gas companies refused to disclose the chemicals in fracking fluid, claiming that the precise mixtures represented proprietary secrets and that offering them up would put the companies at a competitive disadvantage. (Think KFC's secret mixture of eleven herbs and spices, except with as many as five hundred chemical ingredients, including such hard-to-swallow substances as hydrochloric acid, ammonium persulfate, petroleum distillate, naphthalene, and triethanolamine zirconate.[3])

Some of the chemicals that we do know are involved, though, are certainly harmful to human health if they accumulate in our water supplies in sufficient quantities. For example, here's a list of what the Centers for Disease Control and Prevention says can happen within minutes to hours of consuming food or beverages with high levels of benzene: vomiting, stomach irritation, dizziness, sleepiness, convulsions, rapid or irregular heartbeat, death (at very high levels).[4] And that's just the first few hours; long-term exposure can lead to a wide range of serious health problems, including an increased risk of developing leukemia. And that's just one of hundreds of chemicals!

Without solid information on what was sent down the well, it was very difficult to determine the source of any particular instance of water contamination. Even when contamination seemed to have been found, the oil industry and its political allies often insisted that causality could not be proved. That is, just because a stream or wells near fracking activity were contaminated with *something* didn't mean that the fracking had actually caused the contamination.

In spite of these sorts of obstacles, it is now beyond dispute that there have, in fact, been cases of water contamination as a result of fracking and related oil and gas development activities. For example, after lengthy legal battles, the Pennsylvania Department of Environmental Protection now maintains a list of "water supply determination letters"; these are cases in which the department "determined that a private water supply was impacted by oil and gas activities."[5] In other words, drilling for oil and gas contaminated a water well.

Importantly, the list includes cases in which "operations associated with both conventional and unconventional drilling activities" resulted in contamination. "Conventional drilling" refers to wells that were not fracked and did not use horizontal drilling; unconventional drilling is the opposite, making up the vast bulk of new wells in Pennsylvania and elsewhere. As of March 2016, the list had 278 entries.

And that's just in one state! Other states have been reluctant to release or even tally up such problems. But academics have also chimed in, with several peer-reviewed studies showing specific contamination events related to fracking. For example, one study published in the *Proceedings of the National Academy of Sciences* in 2014 used a technique based on hydrocarbon tracers to determine the source of methane (the primary ingredient in natural gas) found in water wells.[6] Though methane can indeed occur naturally in water supplies—a common refrain among proponents of the fracking boom—the researchers determined that the particular methane they had detected did, in fact, enter the water supply as a result of gas drilling. In fact, they pin-

THE LITERAL NITPICK · 117

pointed the specific route of access: leaks in the cement and casings of the well, rather than seepage from far belowground.

Other studies have found other types of evidence. In one case, researchers from the US Geological Survey and the US Fish and Wildlife Service found that "hydraulic fracturing fluids are believed to be the cause of the widespread death or distress of aquatic species in Kentucky's Acorn Fork, after spilling from nearby natural gas well sites."[7] This wasn't particularly hard to observe, actually: after a fracking-fluid spill in 2007, the researchers simply watched a significant die-off of species including the blackside dace (a type of minnow), the creek chub, and the green sunfish.[8]

Another study, published in 2013 in *Environmental Science & Technology*, compared the water quality of about a hundred private drinking-water wells—some very close to (within 3 kilometers of) active natural gas wells in the Barnett Shale formation in North Texas, and others farther from gas activity. They found levels of arsenic, selenium, strontium, and barium that exceeded the maximum levels set by the EPA at many of the water wells that were close to fracking operations, and lower levels at the wells farther from gas activity.[9]

There are other studies and examples of water contamination as well; this is not a case in which there are only one or two isolated incidents and the vast bulk of evidence suggests that fracking is safe for water supplies. In fact, soon after Inhofe's statements, the EPA released a long-awaited report summarizing the state of knowledge on fracking and water contamination, and found multiple specific instances in which various mechanisms had led to water supply contamination.[10] Muddying the waters was the caveat that the study did *not* find these impacts to be "widespread" or "systemic"—a tidbit Inhofe and others jumped on to claim victory.

A reasonable person, in trying to honestly assess the risks to water supplies from drilling for oil and gas, would look at all the various routes of contamination, and examine all the incident reports and spills and faulty cement casings. A politician engaged in the LITERAL NITPICK, though, would not. That politician would instead focus on a single, isolated process, and a single, isolated route of contamination, and would then claim that there is no evidence of any sort of problem. This is what Jim Inhofe did.

Back to the senator's actual statement: "Since 1949, my state of Oklahoma has led the way on hydraulic fracturing regulations, and just like the rest of the nation, we have yet to see an instance of ground water contamination." After the various examples and studies we've just considered, that doesn't sound like it could be true, does it? Inhofe included the entire nation in his claim, and he said there has not been even *one* instance of contamination. What about all the instances we just examined?

The key is in the terminology. Inhofe specifically cited "hydraulic fracturing"—which, technically, means the actual breaking of the rocks a mile belowground by injecting water, sand, and fracking fluid. It doesn't include the drilling of the well, it doesn't include extraction of the gas from those broken rocks, and it certainly doesn't include all the activities taking place on the surface—trucks coming and going, wastewater being hauled away or injected to yet another well, fracking fluid being transported to or from the site, and so on.

Sneaky, isn't it? First, let's agree that Inhofe is correct about

there being no evidence of contamination caused by the fracking process itself. That is, the breaking of rocks has not led to
fluid or methane traveling a mile through the ground to the
water table. How, then, is he wrong?

Take fracking fluid itself; it was this substance that spilled
in Kentucky, causing specific harm to species like the federally
threatened blackside dace.[11] Okay, it was certainly not fracking
per se that caused this harm, but think about it logically: would
fracking fluid have been available to spill if no one was fracking a well? Of course not. All of the various ways that *have*, in
fact, caused contamination exist only in relation to fracking. To
claim, then, that fracking has not caused any problems whatsoever is monumentally misleading.

What's more, even Senator Inhofe's office was willing to con-
cede that his point was remarkably specific. A spokeswoman for
the senator wrote in an e-mail that Inhofe's claim "focuses on
the believe [*sic*] that the physical act of cracking rocks through
hydraulic fracturing, thousands of feet below ground, has never
caused groundwater contamination. What we are not saying is
that a surface spill, faulty casing, bad drilling practices cannot
be a problem."[12]

One might argue that Inhofe is perfectly reasonable to focus
on one part of a process and make a perfectly correct claim
about it. Here's the problem: his point about fracking was made
in opposition to regulations that covered *everything* to do with
drilling oil and gas wells, not just the fracking process itself.
He was not opposing the regulations as they pertained to just
one little thing, but the entire regulation, covering cement well
casings, composition of the fluids involved, information on the

geology of the site, and so on. Inhofe can't have it both ways: either he is concerned *only* with the fracking process, or he is concerned with the entire oil- and gas-drilling milieu.

That's the hallmark of the LITERAL NITPICK: a selective focus on one aspect of an issue, or the strict definitions involved, when logic would require a more expanded view. And with complicated topics like fracking, the technique is remarkably hard for people to see through. Most won't know all the various aspects of gas drilling, and a more general statement like "no instances of contamination" is much easier to understand than a nuanced discussion of which specific processes could benefit from increased regulation and oversight. One way to spot this technique is to look out for universal statements: *zero* instances of contamination. Is that really possible, given all the negative publicity fracking has received through movies like *Gasland*? Could all those opponents be *that* far off?

AT LEAST INHOFE HAD a solid defense: the actual act of fracking likely has not contaminated water supplies. In other cases of the LITERAL NITPICK, though, even the narrowest interpretation doesn't hold water.

In November 2015, acting head of the Drug Enforcement Administration Chuck Rosenberg threw a bit of a tantrum with regard to medical marijuana:

> What really bothers me is the notion that marijuana is also medicinal—because it's not. We can have an intellectually honest debate about whether we should legal-

ize something that is bad and dangerous, but don't call it
medicine—that is a joke. . . .

There are pieces of marijuana—extracts or constituents or component parts—that have great promise. But
if you talk about smoking the leaf of marijuana—which
is what people are talking about when they talk about
medicinal marijuana—it has never been shown to be safe
or effective as a medicine.[13]

Rosenberg, who was appointed by President Obama's attorney
general, Loretta Lynch, argues that "smoking the leaf of marijuana" is at the crux of the ongoing debate about the medicinal
properties of the drug. This is sort of a bizarre misrepresentation of a very wide-ranging discussion: the push to use marijuana as medicine covers FDA-approved synthetic versions of
THC[14] (marijuana's active ingredient), inhaled marijuana, and
other forms. To claim that "what people are talking about" is
solely smoking marijuana is wildly reductive.

The idea that there is no evidence for smoked marijuana's
medicinal benefit also happens to be flat-out wrong. To be sure,
there is not an abundance of evidence, thanks to the drug's
"Schedule 1" status in the United States.[15] That designation
places marijuana alongside dangerous substances such as heroin and LSD (believe it or not), and severely restricts thorough
study of it. The few studies that have been done, however, hint
at potential benefits in a wide range of medical conditions.

For example, one study published in the journal *Neurology* in
2007 found that smoked cannabis reduced pain in HIV-positive
patients with a condition called HIV-associated sensory neu-

ropathy.[16] After five days, patients who smoked the drug had a 34 percent reduction in daily pain, compared with a 17 percent reduction in those who smoked identical cigarettes that were filled with a placebo instead of cannabis. Another study a year later confirmed the effect for this painful condition. A higher proportion of marijuana smokers than placebo smokers again experienced a big improvement in pain reduction.[17]

In one more example, a study from 2012 tested smoked cannabis against a placebo in patients with multiple sclerosis and spasticity, a type of involuntary muscle spasm experienced by many patients. The drug did, in fact, reduce spasticity far better than placebo, and it added in a much bigger reduction in pain as well.[18]

There are not, unfortunately, all that many more of this type of study, so it is hard to make broad, sweeping claims about smoked marijuana's medicinal properties—as some proponents are wont to do. Rosenberg took advantage of this paucity of research to engage in a careful LITERAL NITPICK: by zooming in on the gaps in evidence created by the drug's status, he pushed a potentially useful therapy further toward the margins.

Clearly, the DEA's primary mission to police the country's illegal drug activity doesn't mesh well with the concept of using one of those drugs for medical purposes. But science is far messier than a simple list of controlled substances—the odds that almost anything is as cut-and-dried as Senator Inhofe or DEA head Rosenberg claimed are slim. And once again, these missteps have real-world consequences. Allowing fracking to continue unabated if it *does* cause illness through water contamination is obviously a horrendous outcome, especially given how widespread the natural gas boom has become. And it is a particularly hard-hearted individual who would deprive a dying cancer patient of a bong hit if it would ease the pain of his last days on Earth. To fight back against the LITERAL NITPICK, look carefully at word choice, as well as the specifics and details of the argument. If it sounds like a politician is focusing a bit too closely on one specific aspect, ask yourself: Who threw the baseball?

CHAPTER 8

The Credit Snatch

A HUSBAND AND WIFE LIKE TO TRADE OFF HOUSEHOLD chores every now and then. From March through June, she does the gardening, planting a whole smorgasbord of vegetables. In July, he takes over. After a week, he proudly marches inside with a basket full of tomatoes, squash, green beans, and eggplant.

The dinner guests arrive that Saturday night and dig in to a meal filled with these locally grown delicacies. The husband crows: "I grew them all myself!" And then he sleeps on the couch.

In some scientific fields, just as in gardening, there is often a time lag between when seeds are planted and when the resulting bounty is harvested. Or, even if the timing is shorter, the exact underlying reasons for the harvest may be a bit unclear. That lack of clarity is a politician's best friend.

The CREDIT SNATCH is a ploy whereby politicians claim some sort of accomplishment just because it happened "on their watch." They don't bother to explain the underlying, actual reasons for that accomplishment, which often is a result of a previous administration's policies, or simply has some mechanism behind it that in no way is a result of the elected official's specific actions.

Let's start with former Texas governor and occasional presidential candidate Rick Perry. In a speech in February 2015, Perry listed a litany of accomplishments related to emissions of various types of pollutants over a period when both population and jobs were on the rise (and he managed to get in a jab or two highlighting his climate science denial as well):

> During that same [seven-year] period of time using thoughtful, incentive-based regulation, we decreased our nitrogen oxide levels—which, by the way, is a real pollutant, it's a real emission—nitrogen oxide levels were down by 62½ percent, ozone levels were down by 23 percent, sulfur dioxide levels down by 50 percent, and our CO_2 levels were down—whether you believe in this whole concept of climate change or not—CO_2 levels were down by 9 percent in that state. Isn't that the goal of what we were working towards? . . . We put policies in place that helped remove old dirty-burning diesel engines from the fleets. We were able to transition our electrical power system to the natural gas burning.[1]

That's a lot of numbers, a lot of molecules, and a lot of odd grammatical errors. All those reductions certainly sound good, and in fact Perry was correct—more or less, depending on which years he meant to include—on most (but not all) of them. The questions are, *why* did they happen, and just how impressive are these reductions?

Let's take a look at nitrogen oxide first, the one pollutant reduction for which Perry's number is actually quite misleading. Nitrogen oxides, referred to as NO_x, are an entire family of highly reactive gases that form when fossil fuels are burned. That means NO_x is emitted from car and truck tailpipes, from factories, and of course from power plants burning coal or natural gas.

The NO_x reduction that Perry mentioned (62.5 percent) is a sort of homemade statistic by the Texas Commission on Environmental Quality, and it includes only emissions from "point sources"—that is, major emitters like factory smokestacks and power plants.[2] Those are a big source of NO_x, but so are "mobile sources"—cars and trucks. In fact, mobile sources account for more than half of Texas NO_x emissions, and point sources account for only about one-quarter.[3]

This is notable because Perry specifically mentioned policies that had an effect on "the fleets"—mobile sources of emissions. Some of those policies certainly did help clean up the air a little bit. For example, the Diesel Emissions Reduction Incentive is estimated to cut about 161,000 tons of NO_x over the lifetime of the vehicles it affects in Texas. In 2014, the program resulted in about 54 tons fewer emissions every day, or 20,000 for the year. That sounds great, until you realize that the state's mobile-source NO_x emissions topped 700,000 tons in 2011. So Perry's policies may have put a tiny dent in total emissions, but the reduction he quoted had nothing to do with those policies.

Next up, sulfur dioxide. This noxious gas is produced almost entirely by power plants and large industrial facilities. It may sound familiar, as it is the primary culprit in the formation of acid rain, and the topic involved with Ronald Reagan's founding "not a scientist" pronouncement. Perry is correct about the 50 percent reduction, which again sounds great . . . until you look at the entire SO_2 trend for the United States. Though the years Perry was highlighting are a bit unclear, total US emissions dropped from 11.7 million tons in 2007 all the way to just under 5 million tons in 2014—a reduction of over 57 percent.[4]

The declines in SO_2 and NO_x levels also have almost nothing to do with any of Rick Perry's policies. In fact, these declines have been ongoing for more than two decades, a result of federal policies—exactly the type of federal, top-down regulation that Perry was speechifying against.

In 1990, Congress passed amendments to the Clean Air Act, in large part as a response to concerns about acid rain. The Energy Information Administration, a government institution that collects and analyzes energy and environmental data, had this to say about reductions in emissions:

> The decline in SO_2 and NO_x emissions began soon after enactment of the 1990 Clean Air Act Amendments, which established a national cap-and-trade program for SO_2 and required other controls for NO_x emissions from fossil-fueled electric power plants.[5]

The EIA also attributed some declines to another federal action, the 2005 Clean Air Interstate Rule. Generally speaking, SO_2 reductions have occurred because of strategies implemented by power plants to reduce emissions and meet federally mandated targets. Does that sound like something the Texas governor should be taking credit for?

Finally, Perry decided to deem himself and his administration responsible for a 9 percent reduction in carbon dioxide levels, "whether you believe in this whole concept of climate change or not." Leaving behind the fact that, if one did *not* believe in climate change (meaning one does not understand science in the least), then reducing CO_2 levels would not be an accomplishment to be touted, Perry was again misleading in his claim of credit.

According to EIA figures that were available at the time of Perry's comments, CO_2 emissions in Texas had indeed fallen by 9 percent between the year he took office (2000) and 2011.[6] The only evidence Perry offered for what might have driven down those CO_2 emission levels was the clean-diesel program discussed already, and this: "We were able to transition our electrical power system to the natural gas burning."

We can dispense with both of the arguments in his awkward sentence fragments quickly. First: The transportation sector's CO_2 emissions didn't fall by 9 percent over this time period. They *rose*, by about 7 million metric tons, or about 4 percent.

Okay, what about the electric power sector? Nope, again, CO_2 emissions went up about 4 percent, or about 10 million metric tons. And even if that were not true, Perry's argument about transitioning to natural gas shouldn't really go on his résumé either, since, as we've already seen, the natural gas boom in the United States was a result of improved technologies and economic factors, not some policy magic at the state level.

The decline in carbon dioxide emissions actually came from one particular source: the industrial sector. Manufacturing has declined in general in recent decades in the United States, and emissions have followed suit. In Texas, using the numbers Perry had available, CO_2 from industrial sources declined from about 285 million metric tons in 2000 all the way to 204.6 million metric tons in 2011—a drop of more than 28 percent.

This is an even greater drop than the nationwide trend. According to the EPA, "Greenhouse gas emissions from industry have declined by almost 12% since 1990, while emissions from most other sectors have increased."[7] This sort of decline isn't really the kind of accomplishment most governors are aching to take credit for; in fact, many candidate stump speeches harp on the need to bring back manufacturing jobs, not eliminate them further. And importantly, no one is blaming individual governors for industrial decline; the US State Department's 2014 *Climate Action Report* explains the drop in industrial output and resultant emissions as follows: "This decline is due to structural changes in the US economy (i.e., shifts from a manufacturing-based to a service-based economy), fuel switching, and efficiency improvements."[8]

We've already seen that the electric power sector's emissions actually rose during Perry's tenure, but let's take another look at his claim of a transition toward natural gas, a fossil fuel that emits about half the amount of CO_2 emitted from burning coal. The portion of Texas's power supply coming from natural gas actually *dropped* while Perry was in office, from 67 percent in 2000 to about 61 percent in 2012. Coal use dropped too, from 24 percent to about 21 percent.

The big, fundamental change in Texas electricity had nothing to do with those fossil fuels, but in fact with one source of power: wind. Texas has far and away the most wind power in the country, at almost 18,000 megawatts installed capacity through the end of 2015 (California is second, at just over 6,000 megawatts).[9] That's up from only 184 megawatts at the end of 2000,[10] which took the energy source from less than 1 percent up to well above 10 percent.

Wind power has helped the emissions from the power sector rise *more slowly* than otherwise; they have risen because of a total capacity increase, meaning more fossil fuels have burned even as their proportion of the energy supply has dwindled. Though Perry deserves some credit for helping foster the wind power industry in Texas, the growth had more to do with the potential for profit for landowners and ranchers and the simple fact that the plains of West Texas are an extremely windy place.[11] Moreover, the state's Renewable Portfolio Standard was passed in 1999,[12] before Perry took office.

This litany of misplaced responsibility is a wonderfully thorough example of the CREDIT SNATCH. The politician can claim that all the numbers cited are correct (more or less), and what person listening to the stump speech will bother to check the underlying mechanisms? Emissions fell; that sounds like a good thing, right? It happened on Perry's watch; therefore Perry gets the credit. The effect is that politicians get to claim a track record they don't have. In Perry's case, claiming he was responsible for an improved environment in Texas is particularly galling. Here's how the *Washington Post* described his record as governor (when he still had four more years to keep this up):

He filed a lawsuit against the EPA's greenhouse gas emissions regulations on behalf of the state, a suit widely expected to fail. Perry has said that he prays daily for the EPA rules to be reversed. He has consistently defended oil and coal interests in Texas, notably dubbing the BP oil well blowout an "act of God" and opposing the Obama administration's efforts to regulate offshore drilling in the wake of the disaster. He also fast-tracked environmental permits for a number of coal plants in 2005, cutting in half the normal review period.[13]

And yet there he was, claiming drops in pollution as his own grand creations.

ENVIRONMENTAL ISSUES ARE PARTICULARLY good fodder for a CREDIT SNATCH. Reducing pollution is a catchy and universally popular idea, even if your audience isn't entirely sure what the pollution in question is, and the reasons for those reductions are often nebulous enough that anyone could take credit. "I was there when it happened!" is good enough for many a politician.

Energy supplies are, of course, intimately linked to emissions, and they are another prime target for credit thievery. Here's New Jersey governor Chris Christie, talking at the Iowa Ag Summit in 2015 as he prepared for his presidential run:

In New Jersey, the northeastern states have been part of something called the Regional Greenhouse Gas Initiative, which was essentially a regional cap-and-trade program. When I became governor, I pulled out of that; I don't believe cap-and-trade makes sense. And what's happened in New Jersey? We're now the second largest

solar-producing state in the country, because we've gone
to market-based solutions on helping solar thrive in our
state. And only California produces more solar energy
now than New Jersey.[14]

Christie suggested that removing his state from the RGGI
(pronounced "Reggie"), which is indeed a regional cap-and-
trade program aimed at lowering carbon emissions, somehow
resulted in the solar power boom. Essentially, he took one deci-
sive action—pulling out of RGGI—and claimed it as the reason for this clear example of progress. This is about as drastic
and audacious a CREDIT SNATCH as one can imagine.

First off, the specific claim: solar power is booming in New
Jersey, and this tiny, northern, cloudy state is somehow among
the biggest solar producers in the country. Actually true! Again,
a hallmark of this technique is to base your claim on a true,
verifiable, generally popular accomplishment. Christie checked
that box off perfectly.

According to the Solar Energy Industries Association, New
Jersey had more than 1.5 gigawatts of installed solar capacity
(enough to power about three hundred thousand homes) part-
way through 2015, ranking it third in the country behind sunny
California and *very* sunny Arizona (it was second until relatively recently).[15]

The question is, what exactly has made this cloudy state
that could fit inside California eighteen times over such a solar
powerhouse? The answer has very little to do with Chris Christie. Instead, it has to do with state policies implemented under
previous governors. The state's Clean Energy Program dates to

2001, and the Renewable Portfolio Standard (RPS), which sets a target for how much of a state's power comes from renewable sources, actually was first adopted in 1999 and subsequently updated several times.[16]

Perhaps the most important policy is the Solar Renewable Energy Certificate (SREC) program. Begun in March 2004, the SREC program is essentially a market for the generation of solar power: by producing a certain amount of power from solar (1,000 kilowatt-hours, or a bit more than an average American home uses in one month[17]), the owner earns a single SREC. This credit can be sold for the going rate, generally to utility companies, enabling them to meet requirements set by the state for renewable energy generation.

Most experts directly credit the SREC program, along with the rapidly dropping costs of producing solar panels, with New Jersey's solar boom. Christie only took office in 2010, and though the start of his tenure coincided with a big jump in solar installations, that wasn't his doing, but instead was a reflection of a big jump in SREC prices. A subsequent drop in solar installations in 2013 again mirrored the market, rather than being the consequence of any executive action.

Christie himself has had a mixed relationship with the SREC and RPS programs. Relatively early in his administration, in June 2011, Christie proposed an energy plan that would actually *cut* the RPS, bringing the total of renewable energy required down from 30 percent to 22.5 percent.[18] This was before the solar market peaked and bottomed out, meaning that Christie looked at the booming solar installations and decided to *slow them down*. To his credit, however, when the SREC market plummeted a year later, he reversed course and signed a solar "resurrection bill" designed to address the supply-and-demand imbalance in the SREC market.[19] More recently, Christie vehemently opposed the EPA's Clean Power Plan, a federal regulation designed to cut carbon dioxide emissions from power plants[20]—a move that anyone in favor of expanding renewable energy would not condone.

Of course, in his comments about his energy policy prowess, Christie mentioned none of this. Instead, he focused on his state's removal from RGGI and what he termed a return to "market-based solutions." This makes no sense at all. RGGI, as Christie himself described, is a regional cap-and-trade program. That means it creates a market in which polluters (factories, power plants, and so on) buy and sell emissions allowances in an effort to reduce those emissions. Doesn't that sound remarkably like a market-based solution? It should! Here's the very first sentence on RGGI's website: "The Regional Greenhouse Gas Initiative (RGGI) is the first market-based regulatory program in the United States to reduce greenhouse gas emissions."[21] Christie pulled out of the idea he professes to love so well! And it was working: RGGI estimates that in its first four years of existence, it raised about $1 billion for the states involved and prevented 10 million tons of CO_2 emissions, equivalent to taking nearly two million cars off the roads entirely.[22]

This closer look reveals that New Jersey's impressive solar power progress is based on groundwork laid long before Christie took office, and Christie's tenure may have actually impeded the progress of renewable energy. But why let that stand in the way of the governor's lovely revisionist history?

THE CREDIT SNATCH IS most commonly used by politicians running for election or reelection, as was the case with Perry and Christie. Sometimes, though, it rears its head from politicians on the way out the door. Can you guess who said the following?

> By encouraging cooperative conservation, innovation, and new technologies, my Administration has compiled a strong environmental record.[23]

If you guessed George W. Bush, you're right! A snapshot of his White House website from 2008, at the end of his tenure, shows a couple more specific boasts: US wind power production increased by more than 400 percent since 2001, and solar power capacity doubled between 2000 and 2007.[24] Both sound great, but it is exceptionally difficult to draw a straight line from anything the Bush White House actually did to those increases in wind and solar, or really to any sort of "strong environmental record."

Two factors are primarily responsible for the nationwide growth in renewable energy: technological progress and state policy. State policy means those Renewable Portfolio Standards, largely: a total of twenty-nine states (plus Washington, DC) had an RPS in late 2015, while another eight have a nonbinding renewable energy "goal." Most of the renewable energy in the United States sits in states that encouraged such development with an RPS. This is not a coincidence, and all this growth has little to do with who was president at the time.

The other factor was technological innovation. Advances in the manufacturing of solar cells and wind turbines have driven costs down to a level at which these power sources can compete with coal and natural gas. Here's a Department of Energy chart showing falling costs and subsequently rising deployment of wind power:[25]

DEPLOYMENT AND COST FOR U.S. LAND-BASED WIND 2008–2012

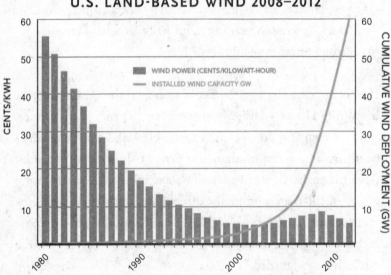

Credit: US Department of Energy

Solar power has had an equally epic price drop over a couple of decades, likely with even more reductions soon to follow. In this case, the federal government actually has played a role in that drop with SunShot, a Department of Energy funding initiative specifically aimed at making solar cost-competitive with other forms of energy. But the initiative was announced in February 2011, long after Bush had left office, and after five of its ten years, the solar industry had made it 70 percent of the way to specific cost goals.[26]

President Bush's actual accomplishments with regard to renewable energy are less than stellar. His 2004 budget, for example, slashed funding for wind power development by 5.5 percent and for geothermal energy by 3.8 percent.[27] In 2006 he doubled down on this approach, cutting the Department of Energy's renewable energy budget by 5.6 percent from the previous year.[28] This happened yet again at the end of his presidency, with his 2008 budget. He signed the Energy Policy Act of 2005—which the *New York Times* said "would shower billions in undeserved tax breaks" on oil and gas companies[29]—only months after saying that cheap oil prices were more than enough incentive for fossil fuel companies.[30] He also opposed subsidizing wind and solar power, along with energy efficiency measures, at various times.[31] Not exactly the record of a solar and wind maven, is it?

The underlying lesson behind the CREDIT SNATCH is one we've already seen: correlation does not equal causation. The simple act of being in office when something improves or happens does not mean that a particular politician's presence *made* it happen. The impact of this technique has a lot to do with its most common deployment, during elections. If voters believe a candidate has a track record on a certain issue that matters to them, it could make them more likely to put that candidate back in office. Once that happens, the lack of true support for the issue in question could rear its head in ugly ways: slowed growth in renewable energy, say, or the lack of any real effort to cut dangerous emissions. When a politician claims credit for a big, long-term trend, do a quick search on what has caused that trend; you just might uncover some credit thievery.

CHAPTER 9

The Certain Uncertainty

A DOCTOR ENTERS AN EXAM ROOM WITH A FOLDER, LOOK-ing grave. "I'm very sorry to tell you this," he says, "but you have been diagnosed with lung cancer."

The patient in the room turns white with fear. "Oh no!" But the doctor quickly offers some reassurance, saying, "Don't worry, we caught it relatively early. Your cancer is stage I, meaning it has stayed small and remains only in your lung and has not spread to other parts of your body. There is a fairly good prognosis with your type of malignancy."

The patient begins to breathe a little easier. "Wow, okay, that's a bit of a relief. What's the treatment? When does it start?"

The doctor gives a condescending guffaw. "Oh no, we're not going to *treat* your cancer. You see, we haven't figured out completely why the cancer formed, and we can't treat it effectively in *every* patient; really, there's still a whole lot of uncertainty around this disease. We wouldn't dare move forward without knowing every last thing about this; doesn't that make the most sense?"

This example sounds ridiculous, but it actually mirrors the same "logic" that is trotted out far too often by our elected officials. No honest scientist would ever say his or her field of study is *finished*, or completely understood. This sounds obvious. We haven't cured Alzheimer's, or traveled at light speed through the galaxy, or figured out exactly how life began. Politicians, though, love to seize on the lack of utter, complete, 100 percent proof in scientific fields as a way of delaying or opposing any action on that topic. Meet the CERTAIN UNCERTAINTY: essentially the doctor's ludicrous argument that since we don't know it all, we don't know anything. As you can probably guess, this tactic has been deployed repeatedly and aggressively in the debate over climate change.

Climate science is inherently a muddy endeavor. Scientists are trying to predict what will happen in the future from a variety of clues, such as long-ago changes to the planet's climate (known as paleoclimatology), computer modeling of what has happened and what will happen, measurements of temperature from many sources, measurements of ice thickness and rate of melting, measurements of sea-level rise and storm intensity and drought and precipitation and so on. The world is a big place, the climate is a complicated beast, and understanding it is not a simple task. But here's the rub: in spite of all those challenges, we know an *incredible* amount about the climate and how it is changing.

The assessments of global warming, including the fact that we have already raised Earth's temperature by about 1°C and that we're on track for a whole lot more in the coming years and decades, are based on very solid science. The fact that humans have caused virtually all of that warming is also a rock-solid conclusion. These are not guesses. To be sure, like any measurement, what we know about warming does have uncertainty associated with it, but that does not make it wrong, nor does it make acting to mitigate it impossible or foolhardy.

To Republican politicians, though, that is exactly what the uncertainty does. They rely on the careful nature of scientists, who are not afraid to admit the lack of total precision, to push for no action at all. There are many examples, but let's start where we left off last chapter, with George W. Bush, speaking during a presidential debate in 2000:

> Look, global warming needs to be taken very seriously, and I take it seriously. But science, there's a lot—there's differing opinions. And before we react, I think it's best to have the full accounting, full understanding of what's taking place.[1]

The idea that we should have a "full understanding" before acting sounds eminently reasonable, which is why politicians have gotten away with this tactic for so long. "Look before you leap" is difficult to argue against, especially to a skeptical public who may not understand exactly what goes into climate science. Before we get into some other more specific claims regarding uncertainty in climate science, here's a quick counter to the entire argument from cognitive scientist and psychologist Stephan Lewandowsky:

So anyone who says that we shouldn't act on climate change because of uncertainty is really inviting you to ride towards a brick wall at 80 km/h because it might not hurt. [Do] you feel lucky?[2]

Well, punk? *Do* ya?

POLITICIANS DEPLOY THE CERTAIN UNCERTAINTY on several specific climate-related topics. As discussed already, climate science is a broad and complicated field, so there are many specific topics on which one can claim a lack of proof. For a first example, let's jump ahead from that debate quote by fifteen years and one Bush brother (Jeb):

I don't think the science is clear of what percentage is man-made and what percentage is natural. I just don't— it's convoluted. And for the people to say the science is decided on this is just really arrogant, to be honest with you.[3]

We'll ignore the fact that doubting what the entirety of the world's scientists say on an issue seems a tad more arrogant than those scientists' explaining what is known. Focus on the message here: the science is not "clear" as to humanity's contribution to warming; the science is "convoluted." By casting doubt on what we know, Bush argued essentially that we know nothing. He was, of course, wrong.

Science does, in fact, have a good idea of how much of the

observed warming is due to human impacts. Spoiler: pretty much all of it.

Here's how the *Fifth Assessment Report* of the Intergovernmental Panel on Climate Change (IPCC) summarized this question in 2013:

> It is extremely likely that human activities caused more than half of the observed increase in [global mean surface temperature] from 1951 to 2010.[4]

The report added that this conclusion is "supported by robust evidence from multiple studies using different methods." In other words, we are really sure about this, and there is a *lot* of evidence for it.

A primary method scientists have for determining what caused a certain climatic phenomenon—in this case, warming temperatures—is called a *fingerprint analysis*, or fingerprint study. Basically, the idea is to look for the unique mark of certain activities on the climate system—an imprint of humans placed on the world. Scientists do this by comparing what we observe to what the computer models would predict.

Models are complicated computer simulations that try to replicate what goes on in the climate. There are many of these in use by scientists around the world. By allowing researchers to add and subtract elements in their predictions, the models make fingerprinting possible. For example, a model might begin from near the start of the Industrial Revolution (say, around the year 1800) and run through the present day. In one scenario, the model simulates what happens to the climate when humans emit all the carbon dioxide that we really have emitted since then; in another, it simulates the same time period but without the CO_2 emissions.

What these studies show is that when you remove human influence from the equation, all the warming we have observed in the real world disappears. No humans, no warming. Here's a chart from NASA[5] illustrating this idea:

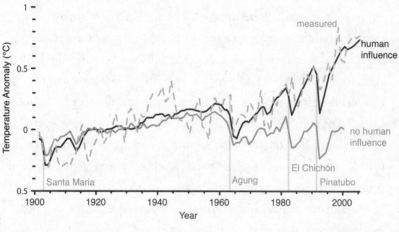

Credit: NASA

The lines representing modeled human influence and observations match up extremely well, while the "no human influence" line stays largely flat and even shows a cooling trend (the chart notes some major volcanic eruptions, which have a cooling effect on global temperatures).

The models don't just look at human CO_2 emissions. They include a wide variety of inputs, such as other emissions, like sulfate aerosols (which have a cooling effect, though a brief one), and nonhuman activities like volcanic eruptions and solar activity. And study after study, of a variety of types, have all pointed to humans as the primary cause of the significant warming observed.

The IPCC has tried to estimate specific temperature changes that can be attributed to various of these causes: human greenhouse gas emissions, solar activity, and so on.[6] Their answer, in short: humans have caused all the warming we have seen, while natural variability and other causes, such as solar output, account for essentially none of it.

These conclusions, as the IPCC noted, come from a wide variety of sources. Here are just a couple of them: one study, published in the journal *Climate Dynamics* in 2012, used a statistical method to assess the various contributions to warming. The authors concluded: "The expected warming due to all human influences since 1950 . . . is very similar to the observed warming."[7] Another, published in *Nature Geoscience* in 2011, combined multiple models to ask similar questions and found that it is "extremely unlikely" that the observed warming could be caused by internal variability—that is, by nonhuman influence.[8]

Again: the world has warmed, and we did it. It's us, and essentially nothing else. Yet many politicians continue to parrot the same general message: we don't know how much the world is actually warming, and we don't know how much warming is due to human activity; therefore, let's not do anything about it.

The last part of that argument, that we should do nothing at all, is its own type of the CERTAIN UNCERTAINTY. Here's Florida senator and 2016 presidential candidate Marco Rubio, appearing on *Face the Nation* in April 2015:

I believe the climate is changing because there's never been a moment where the climate is not changing. The question is, what percentage of that . . . is due to human activity? If we do the things they want us to do, cap-

and-trade, you name it, how much will that change the pace of climate change versus how much will that cost to our economy? Scientists can't tell us what impact it would have on reversing these changes, but I can tell you with certainty, it would have a devastating impact on our economy.[9]

Rubio repeated that same tired refrain, that the human contribution is somehow unknown and unknowable. Then he offered up a particularly brazen version of the CERTAIN UNCERTAINTY: he claimed that scientists "can't tell us" what would happen if we actually tried to slow warming, and said "*with certainty*" that taking *any* actions to mitigate climate change—"cap-and-trade, you name it"—would have a terrible effect on the economy.

Rubio's claim that indeed it *is* certain that taking action would destroy the economy bears scrutiny. This is, obviously, more of an economic discussion than a scientific one, but of course understanding the economic effects of climate change requires an understanding of science. And the experts that have actually undertaken this sort of analysis are clear: doing nothing will cost *far* more than doing something.

The most cited and perhaps most comprehensive analysis of the economics of climate change is now a decade old. The seven-hundred-page "Stern Review,"[10] conducted by UK economist Nicholas Stern and released in 2006, had a stark and firm conclusion: keeping global warming to a reasonable level—1.5°C—would cost money, and a lot of it. In fact, it would cost about 1 percent of the entire world's GDP annually by 2050. But without such actions? The costs would be absolutely devastating, falling somewhere between 5 and 20 percent of global GDP each year.

Other studies have also found huge costs associated with inaction, and surprisingly low costs associated with action. A group at Stanford University led by Mark Z. Jacobson, modeling scenarios in which every state in the United States could transition to 100 percent renewable energy by 2050, has found that such a world would actually cost substantially less than just continuing on our current, fossil fuel–heavy path.[11] Jacobson has said that "when you account for the health and climate costs—as well as the rising price of fossil fuels—wind, water and solar are half the cost of conventional systems. A conversion of this scale would also create jobs, stabilize fuel prices, reduce pollution-related health problems and eliminate emissions from the United States. There is very little downside to a conversion, at least based on this science."[12]

A study published in *Nature* in 2015 found that *not* engaging in such a conversion would have a remarkable effect: a decline in global incomes by 23 percent by 2100.[13] One of the study authors, Marshall Burke, also of Stanford, said: "We're basically throwing money away by not addressing the issue."[14]

Virtually every proposed climate-related policy in the United States is met with that same refrain: that it will destroy the economy. When Rubio made his comments, the primary policy bogeyman was the Clean Power Plan, an Obama- and EPA-led regulation designed to reduce emissions from power plants. Opponents of the regulation, which would have the effect of shuttering some coal-fired power plants and making it nearly impossible to build new ones, claim that the costs would be prohibitive. Indeed, some not-very-independent reports found just such problems. A consulting firm called National Economic

Research Associates released a report in late 2014 that found a total increase in energy costs of $479 billion over the period from 2017 through 2031.[15] Half a trillion dollars!

There are a lot of problems with this report. The most basic appears on the first page, which lists the groups for whom the report was prepared. The list includes the National Mining Association, the American Coalition for Clean Coal Electricity, American Fuel & Petrochemical Manufacturers, and others with more than a little bit of skin in the game. More fundamentally, the analysis ignores a wide variety of factors such as energy efficiency improvements and incentive programs for low-income communities, and it offers very little information about how exactly it arrived at its cost estimates.[16]

If you ask the EPA itself—which, of course, has biases of its own—the benefits of the Clean Power Plan far outweigh the costs.[17] The costs here are technology costs: building new energy infrastructure, retrofitting older plants to reduce emissions, and so on. The benefits are many; for example, by reducing the number of coal plants, a whole host of public health benefits emerge. Along with CO_2, those plants emit other harmful pollutants, such as $PM_{2.5}$, tiny particles that exacerbate asthma, cause other respiratory illnesses, and result in thousands of premature deaths every year. Those outcomes have associated costs, many of which have been quantified.

There are direct climate-related benefits as well, of course; by reducing CO_2 emissions, we slow climate change, meaning we slow sea-level rise, reduce the impacts of stronger storms and lengthier droughts, and so on. These things add up! And perhaps the biggest benefit, though easily the hardest to quantify, is the issue of leadership. The United States was the biggest CO_2 emitter for many decades, and it remains second only to China; a global agreement to slow climate change requires American—and Chinese, and Indian, and Brazilian, and other countries'—leadership. If the United States couldn't even take modest steps to reduce its own emissions, how could it argue for the rest of the world to take such steps? The Clean Power Plan gave the United States a strong leadership position on the issue, a position that arguably helped bring about the international agreement signed in 2015 by 195 countries that will hopefully put the world on a path toward slowing warming.

In the end, the EPA estimated that the net benefits—including the technological costs of updating some infrastructure—would come to somewhere between $26 and $45 billion annually by 2030. Put another way, the EPA calculated a dollar number similar to that estimated by NERA, but in the exact opposite direction, with a benefit instead of a cost. The two sides were, incredibly, a trillion dollars apart in their analyses.

Even CitiGroup (which now goes by just "Citi"), not usually considered an environmental champion, has stated clearly that letting climate change happen is a lot worse economically than the alternative. Citi released a study estimating that an "Action scenario" would cost about $190.2 trillion over twenty-five years—a hefty price tag.[18] But wait! The "Inaction scenario," in which we do nothing but continue burning fossil fuels and continue on our current path, would cost $192 trillion; and that's before the damages due to a warming world are even factored in. Allowing the planet to warm 2.5°C, Citi said, would shave $44 trillion off the global GDP. Letting the increase in temperature go all the way to 4.5°C, which is feasible if we don't stop emitting soon, would cost $72 trillion. That's more than four times the current US GDP, and only $5 trillion less than the *entire world's* GDP in 2014.

The point is this: Rubio claimed that the economic hardships associated with reducing emissions are written in stone. In actuality, they are—but they are opposite from what the senator said. There is near-universal agreement at this point—except among the industries fighting to stay relevant in a changing world, like coal and oil—that acting strongly to mitigate climate change would have a positive economic effect. Rubio's CERTAIN UNCERTAINTY leans on the idea that taking action requires a leap into some big black hole of unknowns. In actuality, the science itself is far more certain than uncertain, and the economic impacts don't even vaguely resemble his assurances.

LEANING ON SUPPOSED UNCERTAINTIES in science is an old trick in political circles. Perhaps the biggest environmental issue of the 1980s, acid rain, lingered for years because of the supposed need for "more study." The Reagan administration consistently pushed back against any action to curb the pollution that causes acid rain, often claiming that the science was simply too uncertain to justify action. For example, EPA administrator Lee Thomas—with a background not in science, but in criminal justice and public safety, before being appointed by Reagan[19]—claimed in 1985 that "people want more certainty" that reducing pollution would even have an effect.[20] He made the same argument again in 1986:

> We do not believe that the current state of [scientific] knowledge can sustain any judgment with respect to the level of emission reductions needed to prevent or eliminate [acid rain] damage. . . . We encourage the Congress to postpone a decision on creating a national acid rain program until it can be based on . . . sounder scientific foundation.[21]

The argument is clearly recognizable: we don't know everything; therefore we shouldn't do anything. And by the time of those comments, Thomas was completely incorrect: we had substantial understanding of how acid rain forms, what it does to the environment, and how reducing emissions would reduce those impacts. For example, a report prepared for the White House Office of Science and Technology Policy and released in 1984, despite accusations of malfeasance on the part of those trying to delay action,[22] found a direct relationship between sulfur dioxide pollution from smokestacks and acid rain.

Even more notable was a report released by the National Academy of Sciences in 1983, which found the following: "If emissions of sulfur dioxide from all sources in this region were reduced by the same fraction, the result would be a corresponding fractional reduction in deposition."[23] In other words, cut the pollution, cut the acid rain. It was essentially settled science several years before the Reagan administration's continued use of the CERTAIN UNCERTAINTY.

THOUGH THE CERTAIN UNCERTAINTY is most often deployed with climate change and related issues, it does rear its head in other scientific arenas as well. For example, in early 2007, Texas governor Rick Perry issued a mandate requiring all sixth-grade girls to receive the human papillomavirus (HPV) vaccine. A few months later, an angry Texas state legislature shot down the governor's plan, passing a bill in convincing fashion that barred the state from instituting such a mandate. One of the bill's sponsors was Representative Dennis H. Bonnen; here's what he had to say about requiring the HPV vaccine:

> We did not want to be the first in offering young girls for the experiment to see if this vaccine is effective or not.[24]

As usual with this type of scientific trickery, the statement sounds eminently reasonable. If we really don't know that this vaccine is effective, why on earth would we *require* it of hundreds of thousands of children?

Because we *do* know that it is effective. At the time, the only HPV vaccine available was Gardasil (others have since been developed and approved). The idea behind the vaccine is to prevent transmission of certain types of HPV (there are dozens of variants of the virus), because HPV is known to be a primary cause of cervical and other cancers. The virus is exceptionally common; the CDC says that "nearly all sexually active men and women get it at some point in their lives."[25] Often the infection is not noticed and dissipates without treatment. Certain dangerous strains, though, raise the risk of cancer dramatically. If the vaccine works and reduces the transmission of HPV, tens of thousands of cancer deaths could be avoided every year.

Now, there is a problem with evaluating a vaccine that prevents a viral infection that can then cause cancer: time lag. The vaccine would ideally be given before an individual becomes sexually active—hence Governor Perry's mandate regarding sixth-grade girls, at a point when the vast majority have yet to become sexually active—but that means years before an actual signal regarding cancer prevention would become clear. The cancers caused by HPV tend to start cropping up when patients are in their twenties at the earliest, meaning as long as two decades between vaccine administration and any indication of whether cancer was prevented.

That does not mean, though, that we don't know whether the vaccine works. It actually isn't very hard to test whether Gardasil prevents HPV itself, and there is also a cancer precursor known as cervical intraepithelial neoplasia, or CIN, that occurs well before the cancers themselves form. And the evidence, even back in 2007, was clear: vaccinating young people works.

Scientists from the pharmaceutical company Merck (which produces Gardasil) presented results of one large trial—the FUTURE II study—in October 2005.[26] That study, including more than twelve thousand young women aged fifteen to twenty-six, found an impressively high rate of effectiveness for Gardasil. In the group that received the vaccine, there was only one case of CIN; in the group that received a placebo, there were forty-two such cases, and one case of cervical cancer. Those results add up to a vaccine efficacy rate of 98 percent.[27]

There was also evidence specifically in younger girls. Another study presented in 2005 showed that girls aged ten to fourteen responded very well to another very similar vaccine. The study resulted in 100 percent of the girls testing positive for antibodies to HPV 16 and 18, the two most important strains of the virus.[28] The presence of the antibodies means that the virus would almost certainly not be able to take hold in those girls.[29]

Soon after Gardasil was approved by the US Food and Drug Administration in June 2006, a federal vaccine advisory panel unanimously voted to recommend vaccination in all girls and women aged eleven to twenty-six.[30] This expert endorsement essentially meant that the community of physicians who know best thought that Governor Perry's mandate was an excellent idea. They believed that there was ample evidence for young girls to receive this vaccine, and that it would undoubtedly save lives. But the Texas state legislature claimed that administering the HPV vaccine was tantamount to experimenting on young girls.

The science was clear back then, and it has only become clearer since: *all* young people, both boys and girls, should receive the HPV vaccine before becoming sexually active, according to the CDC.[31] In the end, the controversy over HPV vaccination had more to do with arguments about sexuality and morality and the like than it did with science. Representative Bonnen's claim of a lack of knowledge was simply a way to support his ideology. This is another hallmark of the CERTAIN UNCERTAINTY. When you hear a politician use this technique—calling for more research before action is taken, or claiming that the science is unsettled—look at the background for that claim. Does the politician support certain special interests where focusing on uncertainty would help push his or her views? Is an oil company–friendly policy or an abstinence-only puritan ideology at the root of the claim?

Politicians break out the CERTAIN UNCERTAINTY only when it suits them. Remember in Chapter 1, when a long line of members of Congress repeated some version of the idea that the "science is settled" as it pertains to fetal pain? With regard to scientific issues, elected officials pick and choose when to argue that uncertainty is both present and important, and the decision rarely has anything to do with the actual state of the research in that field. It is political opportunism and little else. Science will always have uncertainty. Every measurement, every experiment, every theory and hypothesis, including those proven to within the absolute limits of science and understanding—they all have some margin for error. Pointing that out when it suits a political agenda isn't an argument; it's just a smoke screen.

CHAPTER 10
....

The Blind Eye
to Follow-Up

Consider the following quotation, from a speech introducing the winner of a Nobel Prize, with certain identifying words removed:

> [REDACTED], despite certain limitations . . . must be considered one of the most important discoveries ever made in [REDACTED], because through its use a great number of suffering people and total invalids have recovered and have been . . . rehabilitated.[1]

A Nobel Prize and fawning description of a therapy so revolutionary that it gets a "most important ever" designation within its field—what amazing treatment could this be? The advent of chemotherapy? The polio vaccine? *Penicillin?*

No; as a matter of fact, the groundbreaking therapy in question was: frontal lobotomy. The two bits redacted from the quote are in place of, respectively, "frontal leucotomy," which is another term for lobotomy, and "psychiatric therapy." This is not a joke: a Portuguese surgeon named Egas Moniz won the 1949 Nobel Prize in Physiology or Medicine for the idea of hammering into your brain to cut some of its connections.

The lesson here is that science and medicine are never finished. Just because something appears true in 1949 does not mean it will appear true in 1950, let alone in 1985 or 2000 or 2015. Politicians, though, sometimes ignore this basic tenet: this is the BLIND EYE TO FOLLOW-UP, when an elected official repeats information from what may be outdated, improved-on, or outright debunked scientific inquiry.

Of course, it wasn't long after Moniz's prize that the use of lobotomy began to decline dramatically. This drop-off was due in part to the development of antipsychotic medications and other treatments that could help some of the people previously deemed lobotomy-worthy, and in part to a growing and fierce opposition to the use of "psychosurgery" as a method of institutional control. This may be a particularly strange and stark change in medical practice, but this sort of transformation takes place all the time. Scientists and doctors can't simply stop paying attention to the march of progress; the science moves forward, and so do we.

For politicians, though, sometimes the purpose of mentioning a certain scientific tidbit is to advance a political agenda. Therefore, the *previous* study or finding, or one from five or ten years ago, might be a more convincing data point than whatever has cropped up since then. For example, here's what President Obama had to say as he touted the brand-new "Precision Medicine Initiative," specifically speaking of the incredible benefits gained from the Human Genome Project:

One study found that every dollar we spent to map the human genome has already returned $140 to our economy.[2]

A $140-to-$1 return on investment! That is a truly remarkable boon to humanity and to science. And the president was correct: there was indeed a study that arrived at that number. But there was another study afterward, by the very same researchers, that lowered that enormous ROI down to $65 for every dollar spent on the HGP. Obama chose to mention the higher number, ignoring the second study.

First of all, let's be clear: literally no honest scientist on the planet would argue that the Human Genome Project, which took a decade or so to map out all three billion DNA base pairs that make up our genetic code, was not an amazing and crucially important achievement. Since its completion, scientists have used its findings to learn staggering amounts about our genes, our bodies, the diseases that harm and kill us, therapies to stave off those diseases, and much more. It was truly one of science's greatest accomplishments—and $65 to $1 is still an amazing ROI!

The president's BLIND EYE TO FOLLOW-UP was not, in the grand scheme of things, a particularly damaging error. His basic point, that spending money to learn about underlying biology, disease prevention, and so on brings great returns in the long run, is nearly indisputable. The question is, why not use the best, most recent number available? Using the older result only opens you up to criticism, to claims that the money you're asking for isn't worth what you say it is.

The study Obama referenced was conducted by two

researchers—Simon Tripp and Martin Grueber—at the Battelle Memorial Institute, the world's largest research-and-development nonprofit.[3] They undertook "a quantitative measurement of the direct and indirect economic impacts in the United States derived from actual expenditures of the HGP project and follow-on federal expenditures in major genomic science programs," as well as estimates of the impact of the "genomics and genomics-enabled industry."[4]

First, what did the Human Genome Project cost? The federal government invested a total of $3.8 billion (or $5.6 billion, using inflation-adjusted 2010 dollars) in the endeavor, spread out over the thirteen years from 1990 to 2003, to map out every gene in human DNA.

Now the benefits: According to the Battelle analysis, the HGP directly and indirectly resulted in an economic output of $796 billion, along with 3.8 million "job-years" of employment (a job-year is one person employed full-time for one year). That dollar amount comes from a variety of specific sources. Here's how the researchers described them:

> Scientists are using the reference genome, the knowledge of genome structure, and the data resulting from the HGP as the foundation for fundamental advancements in medicine and science with the goals of preventing, diagnosing, and treating human disease. Also, while foundational to the understanding of human biological systems, the knowledge and advancements embodied in the human genome sequencing, and the sequencing of model organisms, are useful beyond human biomedical sciences.

The resulting "genomic revolution" is influencing renewable energy development, industrial biotechnology, agricultural biosciences, veterinary sciences, environmental science, forensic science and homeland security, and advanced studies in zoology, ecology, anthropology and other disciplines.

Simple math based on the expenditures and impacts yields a return on investment of just over $140 for every dollar spent—just as the president said. So why was he wrong?

The president was wrong in choosing the statistic he used because the study authors themselves essentially admitted *they* were wrong. Well, they never came out and said so, but they addressed criticisms[5] by updating the analysis and changing some of their methods—a tacit admission, one could argue, that the first attempt was at least somewhat misleading.

There were several reasons. For example, some of the dollar amounts that were included as benefits are more accurately described as costs; the salaries of the scientists engaged in the project, for example, were initially considered as benefits. Most important, the initial report considered the costs only through 2003, while counting the benefits all the way through 2010. There were, of course, continued expenses associated with genome sequencing and related fields.

When the numbers were run again, both with some changes to methodology and with two more years of accruing economic impacts, the ROI dropped substantially. In the 2013 version of the Battelle analysis, the researchers included impacts and benefits through the same year and concluded: "Every $ of federal HGP and related investment has helped contribute to the generation of an additional $65 in the U.S. economy."[6] That's less than half the impact that President Obama mentioned in his speech, which he gave well after that second report had been released. (He had used the statistic before, notably in one speech announcing the BRAIN Initiative, focused on—obviously— brain research.[7] But that speech happened a couple months *before* the follow-up study came out. He was correct that time, but not two years later. If his staff could find the original study the first time, there is no reason they couldn't find the updated version as well.)

The change to methodology in the two Battelle reports is not any sort of scandal or unseemly incident. Assessing the economic impact of such a broad scientific program is remarkably difficult: what exactly *should* be counted as a benefit related to mapping out the genome? The groundwork laid by the HGP helped scientists discover certain genes that—just for example—play a large role in some forms of skin cancer, which then helped pharmaceutical companies develop targeted therapies specific to those cases, which have helped those people live a little bit longer than in the past. How do we quantify *that*? Each patient who now lives a year longer is an economic boon—ignoring the obvious impact on those individuals' and their families' personal lives, of course—but what about the money spent by the pharma company to develop the drug? The salaries of *those* researchers, not paid by government? And so on; summing up the benefits and costs is an immense and not particularly standardized challenge.

One science policy analyst at the University of Georgia, Barry Bozeman, described this issue to *Nature*: "Coming up with a valid number for the impacts of investments in programs that are extremely complex with extremely diffuse impacts—it's really, really daunting."[8]

In other words, the researchers can indeed be forgiven for their $140-to-$1 number originally quoted, just as the Nobel Prize committee can't be faulted for what was still a general scientific consensus at the time of Moniz's lobotomy prize. But politicians who *use* the science—they should be careful to use the best available consensus at the present time.

In the specific case of Obama and the HGP, his looseness with numbers really could have an impact: the president was asking Congress to fund his medical research initiative, and though its goals were likely universally agreed upon—cure disease and so on—the specifics of how much money to spend certainly were not. If notoriously stingy Congress and its members could point out that Obama seemed to have an overinflated sense of the return on these scientific investments, that could stand as an argument to withhold funding. Getting the science right is always the best bet.

SOMETIMES THE BLIND EYE TO FOLLOW-UP is based on ignorance—willful or otherwise—of a second study, a newer version of older science. And sometimes it is based on wholesale ignoring of an enormous, widely reported topic. Enter the Climategate "scandal."

First, a quick history on this controversy: In November 2009, a still-unknown attacker hacked a server at the Climatic Research Unit, or CRU, at the University of East Anglia in the United Kingdom. This server contained thousands of e-mails sent between climate scientists at CRU and other parties around the world, which thus became public. Climate deniers began sorting through these e-mails and pulling out some choice, totally-out-of-context quotations that they claimed blew the entire global warming "conspiracy" sky-high. They argued that there was clear evidence of tampering with data, of collusion between scientists to report higher amounts of warming than really existed, and so on. All of that was utterly and completely false.

The uproar was truly uproarious; there turned out to be a whole lot of smoke but no fire at all in the CRU e-mails. At least eight separate investigations looked into the entire issue or into certain individuals' roles in the scandal, and *none* found even a hint of wrongdoing. This is where the BLIND EYE TO FOLLOW-UP comes in: before those investigations, an honest politician could perhaps mention Climategate with concern about how science was being done, but that honesty should have led the same politicians to dismiss the entire ordeal as a nonissue once the investigations were completed. Such reasoned thinking is, of course, asking too much of some of our elected officials.

Let's return to an interview given by Alabama representative Gary Palmer, whom we met in Chapter 5 (BLAME THE BLOGGER) arguing that some guy on the Internet had blown open the "biggest science scandal ever." Palmer, while primarily highlighting those new totally false and misleading reports about temperature data "manipulation," also managed to get in a dig about Climategate:

I mean, I wrote about this a couple of years ago, when it came out that the scientists at East Anglia University in England had done this, and that was the data that the United Nations report was based on. It was a huge scandal,

there were emails going around where they were, the scientists were literally talking about how they were going
to change the data. We are building an entire agenda on
falsified data that will have an enormous impact on the
economy.[9]

Okay, Palmer said he "wrote about this a couple of years ago";
had he not read any news since then? If he had, he might have
seen that every investigation, from the Environmental Protection Agency to the US Department of Commerce's Inspector
General[10] to others in the United Kingdom, refuted his telling
of the story.

Here is an excerpt from one such UK investigation, infor-
mally known as the Oxburgh Report, comprising an international panel of experts assembled by the University of East
Anglia itself and by the Royal Society, Britain's answer to the
US National Academy of Sciences:

We saw no evidence of any deliberate scientific malprac-
tice in any of the work of the Climatic Research Unit
and had it been there we believe that it is likely that we
would have detected it. Rather we found a small group
of dedicated if slightly disorganized researchers who
were ill-prepared for being the focus of public attention.
As with many small research groups their internal proce-
dures were rather informal.[11]

Back in the United States, here's how the EPA characterized
its findings, specifically in response to deniers' petition to recon-
sider a finding allowing the agency to regulate greenhouse gases:

Petitioners say that emails disclosed from CRU provide evidence of a conspiracy to manipulate data. The media coverage after the emails were released was based on email statements quoted out of context and on unsubstantiated theories of conspiracy. The CRU emails do not show either that the science is flawed or that the scientific process has been compromised. EPA carefully reviewed the CRU emails and found no indication of improper data manipulation or misrepresentation of results.[12]

The other reports found essentially the same thing: literally no independent investigation has concluded there were nefarious goings-on of any kind hidden among the CRU e-mails. These reports—all of them—were released years ago, in 2010, 2011, and so on. And yet there was Palmer, lamenting "an entire agenda [built] on falsified data."

In some ways, the BLIND EYE TO FOLLOW-UP is among the easier errors in this book to cut through, especially when the topic in question is as widely covered as Climategate. Just read the news! Any consumer of current affairs would have had at least an inkling that Palmer was a bit off, and a quick Google search could confirm that. The challenge, though, arises when it isn't just one lone congressperson out there, shouting into the void. Yes, Palmer was not alone.

Though many politicians continued to raise the Climategate specter long after the exonerations had begun flying around like confetti, one in particular seems loath to let it go: Senator James Inhofe of Oklahoma, climate TOAD-in-chief, author of—seriously—a book titled *The Greatest Hoax*.

Here's Senator Inhofe in January 2014:

Then of course in '09 when ClimateGate came, people realized the United Nations committee, the IPCC, had rigged the science on this thing. . . . Now they're trying to say this cold thing we're going through is just a bump in the new climate. That isn't true at all. It is a hoax.[13]

Of course, "this cold thing" the senator mentions could also be described as "January," or perhaps "winter." And Inhofe has not stopped beating this drum; in the summer of 2015 he again claimed that "Climategate should have ended it right there at that time."[14]

With most of the errors or missteps we have covered, you'll notice no attempt to assign intent. How can we know exactly why these elected officials get science wrong in various ways? Sometimes it may be real ignorance or misunderstanding. After all, science is hard for everyone, be they sanitation workers or senators. It requires a remarkable degree of leniency, though, not to see this particular brand of the BLIND EYE TO FOLLOW-UP as intentionally misleading; these officials, Inhofe in particular, are undoubtedly aware of the reports that upended the claims of malfeasance at CRU and elsewhere. They are, it seems, actively choosing to ignore those reports in order to advance their particular political agenda—an agenda that involves blocking any and all action to mitigate climate change. And of course, this sort of obfuscation of an issue has undoubtedly had real impacts; the ministrations of Inhofe and his ilk in the US government played a large role in preventing national and international action on climate change for decades. It took until the twenty-first international meeting in Paris in late 2015 before the world finally managed to unite toward meaningful climate goals, and political meddling of the type seen by Inhofe and Palmer is at the root of that long delay.

FOR A FINAL EXAMPLE of the BLIND EYE TO FOLLOW-UP, we turn to the Frankenfish. More properly known as the AquAdvantage salmon, this genetically engineered (GE) fish grows into a full-sized salmon in half the time as its naturally occurring counterparts. That means less money spent on feeding them, and quicker access to the food source in general. And it means dramatic and angry attacks from politicians.

The salmon is the first-ever example of a genetically modified animal ("GMOs," genetically modified organisms, is the blanket term for plants and animals made in this way) that the US Food and Drug Administration has approved for consumption. GMOs are a remarkably contentious topic among the public and in Washington, and one that breaks along different party and sociodemographic lines than most others. The most fundamental of arguments against GMOs—which are engineered to do such things as increase crop tolerance to drought or herbicides, to improve nutritional value, or for other benefits—is that we do not know whether these foods are safe for people to eat. If we have fundamentally altered the genes of these plants and animals, who knows whether those changes might affect us somehow?

Though this is a reasonable starting concern, the fact is that there are now decades' worth of evidence on a variety of GMO foods, from grains and papayas to the corn you likely consume every day in various forms—essentially all of them showing no harm at all.[15] Politicians who oppose their introduction or approval, though, tend to ignore the accumulating evidence.

When the FDA approved the AquAdvantage salmon in November 2015,[16] Alaska's two senators and single House representative joined together in condemning the decision, engaging in a hearty BLIND EYE TO FOLLOW-UP. Here's Senator Lisa Murkowski:

> I am livid at the FDA's announcement to approve genetically engineered "salmon"—what seems to be more science experiment than fish or food.[17]

Congressman Don Young chimed in as well:

> This harebrained decision goes to show that our federal agencies are incapable of using common sense. From the beginning, I've said the FDA's process fails to consider the threats GE fish pose to natural salmon fisheries, including genetic contamination, interbreeding, and direct competition. By embarking on this science experiment, the FDA ignores fundamental risk questions related to our wild fish species and food safety.[18]

Notably, this is an issue that cuts across party lines, with Democrats as well as Republicans often spewing misinformation about GMOs. California Democratic House member Jared Huffman also released a statement condemning the FDA decision, saying the salmon "has no place in our waters or on America's dinner plates." Like Congressman Young, Huffman warned that "the release of a new hybrid animal . . . could pose a danger to our wild salmon populations, damage the ecosystems they live in, and undermine our domestic commercial fisheries."[19]

The concerns raised here—safety for consumption, and effects on natural salmon—are completely reasonable! Well, they *were* reasonable, before the FDA actually studied these things and found the concerns to be unfounded.

The condemnations from Murkowski and Young make it sound as though this were some hasty, rash decision—*"Oh cool, genetically modified salmon; sure, you can market that, go right ahead."* In reality, the company that makes the fish—AquaBounty— began its approval flirtation with the FDA two decades ago.[20] A slow process of studying the fish followed, and the FDA declared it safe in 2010. The agency spent the next five years reviewing objections before finally offering the approval. This did not happen overnight, and there was plenty of time for critics to follow the process and understand the current state of the science.

That current state is essentially this: the Frankenfish is no different from a regular salmon. The FDA compared data on the GE salmon with normal farm-raised Atlantic salmon. They found no differences with regard to some key hormone levels, and determined "that food from AquAdvantage salmon is as safe to eat as food from non-GE Atlantic salmon."[21] The nutritional profile of the fish was also essentially identical. For all intents and purposes, these are normal fish; they just grow faster.

The science behind GMOs has progressed substantially over the last few decades. Though opponents like to claim that we don't know anything about their safety, that is completely false. The approval of the salmon is no "science experiment," as Mur-

If this Atlantic salmon were genetically engineered,
you wouldn't be able to tell.

Credit: CSIRO

kowski claimed; it is the result of a long and arduous process, with the experimentation already in the rearview mirror.

Congressmen Young and Huffman also highlighted an issue that was reasonable to ask about years ago but has been addressed: the question of whether these fast-growing fish could outcompete wild salmon. The company that makes the GE fish is not dumping them into the ocean; they raise them in inland tanks, with multiple layers of protection between the tanks and any possible route to the ocean. That setup is required by the FDA's approval. Not only that, but the fish are engineered to be sterile; as any fan of *Jurassic Park* might tell you, this actually isn't a foolproof plan, but it adds another assurance that these fish are extremely unlikely to have any effect on wild salmon populations.

When GMOs first hit the scene, questions of safety or environmental effect were appropriate—and they still are, for any new animal or plant that may have a different set of effects or concerns. But a blanket refusal to update one's views with no regard for the steady march of science is absurd. In the case of the Alaskan politicians, the source of this stubbornness is relatively easy to uncover: the seafood industry is among the state's most important, contributing 78,500 jobs and $5.8 billion to Alaska,[22] so protecting it from encroachment makes sense. For others, the fear of genetic engineering is a more visceral one, rooted in science fiction and apocalyptic scenarios.

Politicians with obvious agendas, of course, aren't the best judges of science. Here's how the World Health Organization summarizes the current state of affairs:

> GM foods currently available on the international market have passed safety assessments and are not likely to present risks for human health. In addition, no effects on human health have been shown as a result of the consumption of such foods by the general population in the countries where they have been approved.[23]

Other major scientific bodies say essentially the same thing: we should and will keep testing new GMO products, but as of yet, no harm has been found. This is, unfortunately, a perfect situation for the BLIND EYE TO FOLLOW-UP. Any politician can ignore the latest studies and continue spreading the "genetic engineering is scary" line, and it will take diligent, informed citizens to see through it.

The march of progress in science is often slow and steady,

and to keep moving forward we have to pay close attention to each successive step or misstep. If psychiatry had stopped paying attention in 1949, we might still be lobotomizing tens of thousands of troubled and sick individuals around the country and the world. Instead, the scientific community had a prolonged and healthy debate regarding psychosurgery and, in the mid-1950s, largely abandoned lobotomy as it had been practiced in favor of drugs like Thorazine.

Politicians think they can get away with continued support for lobotomy long after that ship has sailed. Don't let them pick and choose which advancements and developments to embrace and which to ignore. If you pay careful attention to their claims and the ideologies and platforms that underlie them, you can help keep science moving forward.

CHAPTER 11

The Lost in Translation

STICKING WITH THE SEAFOOD THEME, A QUESTION: HOW much fish do you eat?

Related question: Are you a pregnant woman who subsists primarily by catching and eating your own fish from a lake?

If you are an average American, your answer to the first question should be "a bit less than five grams per day on average,"[1] or about four 1-pound filets of salmon, tilapia, or trout per year. And if you are an average American, your answer to the second question is almost certainly, "What? No. What the hell kind of question is that?"

If you happened to listen to the remarks of former Pennsylvania senator and multi-attempt Republican presidential candidate Rick Santorum one day in April 2015, you would have heard that the Environmental Protection Agency was using some fairly outlandish answers to those questions in order to justify one of its regulations. Here's what Santorum said:

And here's the calculation [EPA officials] made. The average woman in America consumes five ounces of ocean and
lake fish a week. . . . This is their assumption: that pregnant women in America will consume not five ounces, but
six pounds! Six pounds . . . that they caught themselves.[2]

The regulation that Santorum was railing against is called
the MATS rule—Mercury and Air Toxics Standard. It is aimed
at reducing the amount of mercury emitted from smokestacks,
because mercury, generally speaking, is bad for us; it is especially bad for unborn children, affecting nervous system development and potentially causing deficits in IQ in the children
later in life. But first, Santorum went on, mangling sentences
and statistics along the way:

So women in America, six percent of all women in
America—[the EPA] concentrated in around the Great
Lakes area, so if you're a Minnesotan woman, twenty-
one percent of Minnesota women who are pregnant fish
for six pounds of food a week that they consume. . . .
These fisherwomen who are out there on Lake Superior
catching fish, filleting it and eating it themselves, they're
going to pass on mercury to their children.

Here's a good rule of thumb for evaluating scientific statements of any type at all: if something sounds completely outlandish, it probably is! (Note: Rule does not apply to actual
principles of quantum physics. In that case, outlandish is far
from a disqualifying characteristic.) Does what Santorum said

sound remotely feasible? Would the EPA, no matter your opinion of the agency, actually assume that 6 percent of all American pregnant women or 21 percent of Minnesotan pregnant women catch and eat 6 pounds of fish every week? Those are truly absurd numbers.

Of course, it was not at all feasible; the EPA made no such assumption. The number they actually used was about 0.12 pounds per week—that's *fifty times* less than Santorum claimed. And not only that, but the women in question were not assumed to be subsistence fishers, out there on Lake Superior as Santorum characterized them; the analysis included anyone who lived in a recreational angler household.

How did this happen? The LOST IN TRANSLATION, that's how. This error occurs when politicians hear a scientific claim second-, third-, or fourth-hand and, along the way, the truth of the matter gets lost. It's a game of telephone, except with the presidency or a crucially important health-related regulation at stake. This type of error is a cousin of the BLAME THE BLOGGER, since often this garbled chain of information comes from somewhere online. In this case, though, the information *changes* on its way to the mouths of our elected officials; sometimes that change seems intentional, or perhaps honest mistakes are made. Either way, we end up with some of the most absurd claims politicians can make.

In Santorum's case, let's go back to the original source to try to understand his claim. The EPA releases a Regulatory Impact Analysis, or RIA, for every regulation it proposes. For the MATS rule, this complicated document attempts to lay out what it will cost to follow the regulation—in this case, that means adding technology to smokestacks and power plants that will reduce mercury emissions—as well as what benefits the regulation will yield when fully implemented.

The benefits of reducing mercury are health related; in fact, most of the benefits that the EPA could actually monetize are "co-benefits," results of the reduction in *other* pollutants that would occur if mercury-reducing technology was added to the smokestacks. Among those related pollutants is $PM_{2.5}$ (which we discussed in Chapter 9), tiny particles of stuff that enter our lungs and can cause all sorts of health issues. By reducing $PM_{2.5}$ emissions, we know that we can reduce asthma and heart attacks, lower the number of school and work days missed, and even prevent thousands of premature deaths every year.

The EPA found that the MATS rule would have an annual compliance cost of $9.6 billion, compared to annual benefits of between $37 and $90 billion, depending on the estimation method used.[3] That sounds pretty good! And in fact, the mercury-and-fish issues Santorum attempted to highlight with the EPA's analysis had a laughably negligible effect on those overall numbers. That's because the $PM_{2.5}$ benefits are far and away, in terms of dollar values, the biggest that the EPA could assign, while the agency could estimate only a tiny benefit associated with reducing the mercury itself (which would moderate IQ losses in children): an annual benefit of "$0.004 to $0.006 billion," which most of us would simply call "$4 to $6 million."

In other words, the entire premise behind Santorum's fish fantasy is essentially meaningless in the scheme of this regulation. But if the EPA really did make some outlandish assumptions as an attempt to sell its controversial regulation, that's at least worth a look.

Here's the background on the numbers Santorum mangled: To determine the benefits associated with just mercury emissions reductions, the agency did an analysis involving fish consumption. Mercury, once it drops out of the sky, often falls into lakes and rivers, where it is converted to a new form called methylmercury and can accumulate in the fish that live there. We humans eat those fish, and thus we also eat mercury.

In order to determine how useful it will be to cut down on mercury emissions, the EPA had to figure out just how many people are vulnerable. Since the effects of mercury are most notable in unborn babies, the agency focused on the relevant population: pregnant women who consume fish.

The researchers did this by looking at "recreational angler households."[4] The source was a 2006 National Survey of Fishing, Hunting, and Wildlife-Associated Recreation,[5] and—this is key—included any households "that engaged in freshwater fishing during the year." Did you go fishing even *once* last year? Congratulations, you are a recreational angler!

Recreational anglers are a relatively common breed; 27 percent of Minnesotans, for example, engaged in fishing at least once in a year. Other states ranged much as you might imagine: 18.4 percent in Idaho, 5.1 percent in Massachusetts, 22.8 percent in Wisconsin, and so on. The key point is that the EPA is not focusing on some ultra-rare subset of the population that catches fish in order to survive daily life.

So, how much fish do these pregnant women in recreational angler households consume? The EPA relied on earlier studies on this issue, and clearly stated in the RIA that a mean of 8 grams per day was used. That's 56 grams per week, or 0.12 pounds per week. One-tenth of 1 pound; that's a far cry from

THE LOST IN TRANSLATION · *177*

Santorum's 6 pounds per week. Even further, the EPA's "95th percentile" figure, meaning people who ate more fish than 95 percent of the recreational angler population, was 25 grams per day, still less than half of a single pound of fish per week.

So we've seen the source material, and we've seen the end result. How did recreational anglers eating a minuscule amount of fish turn into subsistence fisherwomen out on Lake Superior surviving on 6 pounds of fish per week? It turns out there were a couple of steps between the EPA and the horse's—sorry, Santorum's—mouth.

Let's go backward from the former senator's version. A spokesperson confirmed that Santorum derived his numbers from an op-ed piece in the *Wall Street Journal*, penned by a lawyer named Brian Potts. In that piece, Potts did indeed write that the EPA claimed that 6 percent of all pregnant women in America, and higher percentages in states like Minnesota, "subsist primarily by catching and eating as much as six pounds of lake fish a week."[6]

Already, we can see some LOST IN TRANSLATION errors: Santorum abandoned the "as much as" language, which essentially meant that Potts was citing the high end of the estimates. Santorum made it sound as though 6 pounds was the average, standard amount. Yet the *Wall Street Journal* piece seems to have been wrong as well, if it even mentioned 6 pounds of fish at all. How did that happen? Let's go back one step further, to a brief filed by the Cato Institute, a libertarian think tank, to the US Supreme Court about the MATS rule; Potts said his analysis for the *WSJ* was based directly on that brief.[7]

The Cato Institute's brief is a long and complicated document, in which the authors argue that the EPA's analysis of the

negative impacts of mercury was flawed. The brief contained
a fairly dramatic mistake, however: Cato based its conclusions
on *two* separate EPA documents but conflated the findings from
those as if they were one and the same.[8]

The first of those documents was not the RIA, but a "Technical
Support Document" on the risks associated with mercury. This
document presented what is known as an "appropriate and necessary" finding: before issuing a regulation, the EPA must determine that the regulation in question is, in fact, appropriate and
necessary. To do this, the agency must show that some group of
people would suffer harm if the regulation did not go through—
in this case, they looked at "high-end" subsistence fishers.

Yes, we have finally found our subsistence fishers, these rug-
ged souls catching and eating fish as their primary means of
sustenance! The EPA noted, though, that these people represent
a vanishingly small proportion of all fish consumers around the
country. But no matter. These, surely, are the folks who consistently consume 6 whopping pounds of fish every week!

Well, no, not really. The average daily consumption *only
among these subsistence fisherwomen* was 39 grams. That's only
about 0.6 pounds per week, or ten times less than we're looking
for. (Importantly, the IQ losses Santorum cited that are at the
root of this issue came from the 0.12-pound-per-week rate; at
0.6 pounds, or even higher among subsistence fishers, the effects
on childhood cognitive development grow even greater.) Only
when we look not just at subsistence fishers on average, but at
the very highest consumption rate even among these people
who are already quite reliant on fish, do we find the number
in question: the 99th percentile of subsistence fishers consumed

373 grams of fish per day, or about 5.75 pounds per week—close enough!

But the Cato Institute didn't differentiate between the "Technical Support Document" analysis and the RIA analysis; they claimed that in the RIA the EPA modeled mercury exposure on children born into the subsistence households. But they were wrong. In the RIA itself, the EPA said: "To identify and estimate the size of [the] exposed population, the benefits analysis focuses on pregnant women in freshwater recreational angler households."[9] Cato either got confused or purposely conflated the two documents in order make the EPA's analysis sound more ridiculous.

To recap, this tidbit of information about fish and mercury started out in EPA documents, morphed into something new in a think tank's Supreme Court brief, was mildly twisted by a lawyer writing an opinion piece for the *Wall Street Journal*, and then utterly mangled one final time by a former senator running for president. The LOST IN TRANSLATION, in spite of the resulting absurdities, can be a bit hard to unravel; after all, Santorum did not cite the *Wall Street Journal* in his speech, Potts's op-ed at the *WSJ* did not cite the Cato Institute, and the Cato Institute claimed it was citing the EPA directly but got the analysis fundamentally wrong (not to mention the fact that it is a rare person who will bother reading an amicus brief filed to the Supreme Court by a think tank). The key point is to look for the bizarre, absurd, or downright crazy-sounding tidbits in a political speech. Sometimes these have a strange history, and tracing them back to the source could help you better understand both the science and the policy in question.

———

FOR WHATEVER REASON, the EPA seems to be involved in a
number of LOST IN TRANSLATION examples. Maybe this is because
the agency produces a litany of complicated scientific reports
in order to issue its regulations or enforce existing rules, and
those rules and regulations have been consistent targets of the
pro-business, antiregulation types in Washington.

In another such example, Kentucky senator and 2016 presidential candidate Rand Paul told a ridiculous-sounding story
about EPA overreach in a speech in June 2015:

> Over 40 years, we now define pollutants as dirt and your
> backyard as a navigable stream. It would be funny if we
> weren't putting people in jail for it. Guy named Robert
> Lucas, down at the southern part of Mississippi, 10 years
> ago was 70 years old. He was put in prison for 10 years.
> He just got out. Ten years without parole. Ten years
> without early release. He was convicted of a RICO con-
> spiracy [under the Racketeer Influenced and Corrupt
> Organizations Act]. RICO's supposed to be something
> you go after gangsters for. You know what his conspir-
> acy was? Conspiracy to put dirt on his own land. We've
> gone crazy.[10]

Indeed, jailing a septuagenarian for "conspiracy to put dirt on
his own land" does sound crazy! If only that version of the tale
were remotely true.

Paul was discussing this case in connection to another EPA
regulation, known as the Clean Water Rule, which specifies the

THE LOST IN TRANSLATION · 181

types of lakes, rivers, and streams that are protected from any dumping of contaminants. The rule caused substantial controversy because of a somewhat opaque definition of "waters of the United States": Opponents of regulation repeatedly claimed that anything down to a puddle in your backyard could be considered a protected waterway if the EPA unilaterally decided as much. On the other side, EPA and environmental groups said those claims were completely overblown, and the rule simply represented a clarification of which waterways the Clean Water Act covers.

Senator Paul's story about Robert Lucas predated the new EPA rule, however; he told a version of it in his 2012 book *Government Bullies*, though in that one he had more space to give some of the details. We'll focus on his shortened stump speech version, since far more people would have heard that than would have read his book. First, let's get the basic facts of the case straight. The man in question did get a ten-year sentence, though he was released early, after about seven years. Strike one for Paul. He also could not have been seventy years old when he went to jail, since he was seventy-five when he was released after those seven years. Strike two! And finally, there were, in fact, no RICO charges at all. Paul strikes out before we even approach the fundamentally wrong description of the case in question.

Paul claimed that this old man was sent to prison for literally putting dirt on his own land. Here's how the EPA described it: "The most significant criminal wetlands case in the history of the Clean Water Act."[11] What gives?

The Lucas case began back in the 1990s. Paul may have gotten his version of the story from any of a collection of small Mississippi newspapers in which aggrieved opinion pieces appeared when Robert Lucas was convicted and went to prison, in 2008. These included the Jackson *Northside Sun* (sample headline: "Bureaucratic Nightmare Lands Three in Prison"); the Starkville *Daily News* ("Tyranny Is Still Alive and Well in Mississippi"); and the Greenwood *Commonwealth* ("EPA Frenzy Illustrates Threat to Landowners").

In the McComb *Enterprise-Journal*, several articles by Wyatt Emmerich (who happens to own that paper) sound a lot like Paul's version of events. Emmerich doesn't necessarily dispute the facts of the case, but simply argues that the EPA shouldn't have the authority to call the land in question a "wetland" when there is no standing water and it sits well above sea level.

Well, that argument doesn't hold water. Here's the EPA's official definition, also used by the Army Corps of Engineers:

> Wetlands are areas that are inundated or saturated by surface or ground water at a frequency and duration sufficient to support, and that under normal circumstances do support, a prevalence of vegetation typically adapted for life in saturated soil conditions. Wetlands generally include swamps, marshes, bogs, and similar areas.[12]

No mention of sea level, and a conspicuous inclusion of "ground water." Just because you don't see a puddle doesn't mean you're not standing in a wetland! In the case in question, the fact that the area qualified as a wetland is important, but it's not all that Paul got wrong.

Robert Lucas, along with his daughter Robbie Lucas Wrigley and an engineer, M. E. Thompson Jr., did not just put dirt in their backyard. They filled in those wetlands covering about 260 acres (that's a big yard!) in order to build low-cost housing units, including roads and driveways, as well as septic systems. This development project went ahead in spite of repeated warnings and cease-and-desist orders.

In fact, the Army Corps of Engineers warned Lucas as far back as 1996 that the property he was developing contained wetlands and thus could not be turned into homes. According to the Department of Justice, which indicted the group (on forty-one counts, convicted for forty) in 2004, when George W. Bush was in office: "The indictment recites a long record of warnings that the Mississippi Department of Health and other regulatory agencies issued to the defendants notifying them of the public health threat they were creating by continuing to install septic systems in saturated soil."[13] Again, a wetland doesn't necessarily mean a swamp; this land was saturated with ground water, and underground septic systems don't mesh with saturated soil.

In spite of all those warnings, as well as cease-and-desist orders from both the Army Corps and the EPA, Lucas went ahead with the development and sale of his housing units, known as Big Hill Acres. Then, according to the Department of Justice, it got predictably gross: "The Big Hill Acres residents have suffered from seasonal flooding and the discharge of sewage from failing septic systems onto the ground around their homes." How far have we come from "put[ting] dirt on his own land"?

Reading Emmerich's newspaper columns—or the other small-town papers' antigovernment screeds—on the case illustrate where Paul may have gotten his watered-down version. Emmerich himself mangled the details of the case, arguing that

Lucas had never been sued by those who bought his housing units (Department of Justice: "The development has been the subject of numerous civil lawsuits by tenants against the developers."), that "nobody knows what a 'wetlands' is" (see the specific definition from the EPA and Army Corps of Engineers presented earlier), and that this was all a case of the federal government deciding arbitrarily to "squash them like bugs."[14]

In his book, Paul repeated many of the details from Emmerich's columns, and in his speech he boiled it all down to a much more pithy sound bite. This is one part LOST IN TRANSLATION, one part BLAME THE BLOGGER, one part OVERSIMPLIFICATION— and taken all together, outrageously misleading. This particular case obviously had a significant effect on the people duped into buying totally inadequate properties, but the larger issue of wetland violations is no small problem. A half-million football fields' worth of wetlands disappear in the United States every year,[15] partially because of coastal impacts, but also in part because of land-use changes like that perpetrated by Robert Lucas.

That loss is not trivial. Here's how the Department of the Interior characterizes wetlands' importance:

Wetlands provide a multitude of ecological, economic and social benefits. They provide habitat for fish, wildlife, and a variety of plants. Wetlands are nurseries for many saltwater and freshwater fishes and shellfish of commercial and recreational importance. Wetlands are also important landscape features because they hold and slowly release flood water and snow melt, recharge groundwater, act as filters to cleanse water of impurities, recycle nutrients, and provide recreational opportunities for millions of people.[16]

No one would argue that putting dirt on one's land should necessarily require a prison term. But the Clean Water Act exists for a reason, and enforcement of regulations is sometimes necessary. Politicians can obviously call out what they consider to be government overreach, but it doesn't help anybody if the details become mangled beyond anything remotely resembling the facts of the case. When it comes to legal issues that have a basis in science—preservation of wetlands, in this case—it becomes both easier for the elected officials to misrepresent the truth and more important to avoid doing so. The only way the water and air stay clean is if we—the constituents of these politicians—do our part, by taking these absurd claims and outlandish tales with a grain of salt.

....

CHAPTER 12

The Straight-Up Fabrication

To INTRODUCE THE FINAL TYPE OF ERROR, HERE IS FORMER
Missouri congressman Todd Akin, answering a question about
whether abortion restrictions should make exceptions for vic-
tims of rape:

> It seems to be, first of all, from what I understand from
> doctors, it's really rare. If it's a legitimate rape, the female
> body has ways to try to shut the whole thing down.[1]

The "legitimate rape" quote became a punch line, a calling
card for the antiscience (and antiwomen, according to many
critics) viewpoints of many political leaders. In a way, it was
fortunate that the line was so ridiculous; pretty much every-
body who heard this bit of gibberish understood it to be just
that right away.

This is a prime example of the STRAIGHT-UP FABRICATION,

a claim about science that has no basis in fact or any sort of reasonable or understandable sourcing. We won't go so far as to call such claims *lies*, since intent is a hard thing to sort out, but these are statements that are plucked from thin air and presented as truth, regardless of exactly why the speaker decided to do the plucking and presenting. This is arguably the most depressing of the errors covered here—politicians simply relying on the idea that the loudest voice is often considered the most correct, whether or not their point is defensible or fact based. It can also be, unfortunately, quite hard to see through; without any real origin for the claims, where does one start to try and cut through them? The only lucky aspect is that these are often—though certainly not always—ridiculous in nature, and your BS detector is likely to sniff them out.

With Akin and legitimate rape, there is not much debunking work to do. It is just pure nonsense. Rape is just as biologically likely to result in pregnancy as consensual intercourse; the body does not know, somehow, that the sperm seeking to fertilize an egg are unwelcome guests in the fallopian tube, that the resulting embryo is unplanned and the result of violence and thus should not implant in the uterus. Biology is biology, whether Todd Akin says so or not.

Though his statement was quickly ridiculed from all corners, and Akin lost the Missouri senatorial election largely because of that ridicule, it is wrong to dismiss it as though it had no negative effect whatsoever. Sometimes, making mistakes about science can have effects that reach into the social sphere; in the years since Akin's gaffe, the pervasive "rape culture" of the United States has become a central issue on college campuses and elsewhere. Some have argued that, by spreading misinformation about how women's bodies actually work, Akin contributed to that culture. For example, here's how Michael Jeffries, writing at the *Atlantic*, described the problem:

> We cannot reduce the ignorance of people like . . . Akin to sound bites or place it in the category of election-season inanity. Their statements are the toxic runoff of our culture's failure to prevent and address sexual violence in all its forms.[2]

Again: science doesn't sit by itself, alone in a lab coat, pondering the mysteries of the universe with little outside influence or consequence. When politicians mistake scientific issues, it can have ripples in our everyday lives.

PERHAPS SOMETHING ABOUT sexuality brings out the worst in politicians. The United States is a country founded by Puritans, after all, and we have had a complicated relationship with sex ever since. Many of our elected officials sometimes sound as if they maintain some of the seventeenth-century opinions or values espoused by our Puritan forefathers, especially when it comes to issues regarding sexual orientation.

In an interview in early 2015, as Dr. Ben Carson was gearing up for his presidential run, he disputed the idea that the progress toward marriage equality mirrored the civil rights battles of the mid-twentieth century. The difference, he said, lay in the fact that people have no control over their race, but they can, in fact, choose whether or not to be gay. CNN's Chris Cuomo asked Carson why he thought that was the case, and Carson responded:

> Because a lot of people who go into prison go into prison straight and when they come out they're gay. So did something happen while they were in there? Ask yourself that question.[3]

This is an absurd position to take. Being gay is a choice because people change their sexual orientation in prison? Though Carson insisted this was true to Cuomo, there is absolutely no evidence for his point.

First of all, ignoring the prison red herring for a moment, what does science actually say about sexual orientation and this idea of choice? In short, it is a complicated issue and one we still don't understand particularly well, but there is a general consensus that the vast majority of us do *not* experience our sexual orientation as a choice. The American Psychological Association describes the situation this way:

> Although much research has examined the possible genetic, hormonal, developmental, social and cultural influences on sexual orientation, no findings have emerged that permit scientists to conclude that sexual orientation is determined by any particular factor or factors. Many think that nature and nurture both play complex roles; most people experience little or no sense of choice about their sexual orientation.[4]

Surveys have borne that idea out. For example, one 2010 study involving 662 individuals found that 88 percent of gay men and 68 percent of lesbians "reported having had no choice at all about their sexual orientation."[5] Those numbers rise to 95 percent and 84 percent when including people who say they had a "small amount of choice" in the matter.

If it is not a choice, then what is it? Sexual orientation almost certainly has a genetic component, as shown in several studies over the last few decades. In 1993, researchers published a paper in *Science* showing that male cousins and maternal uncles of gay men—meaning other males who shared some genes—were themselves more likely to be gay.[6] This study, involving 114 families of homosexual men, found that there was no such increased likelihood among fathers or paternal relatives, suggesting that sex-linked transmission (meaning the genetic information stored on the X or Y chromosome, which determines our gender) may play a role. The authors concluded that there was more than a 99 percent chance "that at least one subtype of male sexual orientation is genetically influenced."

More recently, other researchers have suggested that your orientation could be influenced in utero, by "epigenetic" effects.[7] Essentially, this means that hormone exposure differences while in the womb could play a role in turning on or off certain genes that we all have, and the complicated interaction of those genetic switches could play a role in eventual sexual orientation.

A common argument trotted out *against* the notion of genetic underpinnings to homosexuality is the idea that it creates an evolutionary paradox: since gay men and women are less likely to reproduce, wouldn't a gene that determined their orientation die out with ease as time passed? In fact, studies looking into this question have found a convincing explanation for why this is no paradox at all.

In one such study of 98 homosexual men and 100 heterosexual men and their relatives—totaling more than 4,600 individuals—the maternal relatives of the gay men had a higher "fecundity" than relatives of straight men.[8] Fecundity is essentially a measure of fertility; the study found that relatives of gay men have more children. That could offset the relative lack of offspring from the homosexual men themselves, meaning, again, that the paradox is explained. Another study, published in 2012, found essentially the same thing, concluding that the increased fecundity among relatives "compensat[ed] for the reduced fecundity of homosexuals."[9]

So, science has told us that there is almost certainly some genetic component to sexual orientation, and that almost nobody experiences a large degree of choice in the matter. Carson's argument, though, hinged on what he claimed was the experience of people changing their orientation while in prison. Is there any truth to that?

Nope. Carson said that "a lot of people" go into prison straight and come out gay, which somehow would prove that homosexuality is a choice. In reality, there is almost no research on changes to sexual orientation while incarcerated. In fact, only one study—and apparently only one—asked a question that was even remotely related; this investigation was conducted by researchers at the University of Tennessee at Chattanooga in 2013.

The researchers—Christopher Hensley and Lauren Gibson—

conducted a survey of 142 inmates at a single maximum-security prison. These inmates were asked what their sexual orientation was before incarceration, and how they would characterize it "today" (the survey included many other questions on sexuality and other issues as well). Twenty-four inmates did, in fact, report a change in sexual orientation; most of these changed from straight to bisexual.

Hensley himself has said that this was a very small sample and does not necessarily tell us anything at all about the prison population at large, or about sexual orientation and choice.[10] The study also did not follow up with inmates upon release from prison, which would more closely resemble Carson's tortured point. Since this is as close as we can get, essentially zero actual evidence supports what Carson said.

The prison comment is an example of the STRAIGHT-UP FABRICATION that could actually sound convincing to those listening. Carson was assertive and insistent, as if this was simply a known fact about prisons and choice, when in fact he had pulled this tidbit out of thin air. And once again, this is a scientific error that has echoes in other arenas: the LGBTQ community has, of course, faced discrimination and harassment in virtually all parts of life, and misconstruing what is actually known about sexual orientation makes overcoming this monumental civil rights challenge that much more difficult.

IN THE INTRODUCTION, we traveled back to the Reagan administration to find what may have been the first example of the "I'm not a scientist" refrain. His topic was volcanoes;

thirty years later, former Arkansas governor and presidential candidate Mike Huckabee trotted out a STRAIGHT-UP FABRICATION on that very same topic, though focusing on a different volcanic emission. Here's the Huckster in an interview with Katie Couric in 2015, responding to the question of whether man contributes to global warming:

> He probably does, but a volcano, in one blast, will con-tribute more than 100 years of human activity. So when people are worried about it—you know?[11]

Since the question was about global warming, we can only assume that Huckabee meant this mythical volcano is spewing out massive amounts of carbon dioxide, unlike Reagan's claim on sulfur dioxide. The degree to which Huckabee was wrong about this is truly amazing.

The US Geological Survey, citing various available studies on the topic, says that the average total CO_2 output of *all* the world's volcanoes is 0.26 gigatons—a gigaton is 1 billion metric tons, in case you forgot—per year.[12] Humans, meanwhile, are responsible for about 30 gigatons every year (though hopefully that number will start to come down soon). That means we have volcanoes beat by more than a hundred times.

But wait, it's worse than that! Huckabee said you would need *one hundred* years of human activity—so actually, he was off by ten thousand times, or about *one million percent*.

Except the governor said a *single* volcanic blast, not the entire global population of volcanoes! One of the biggest erup-tions in recent memory was Mount Pinatubo in the Philippines in 1991. According to the USGS, that eruption released 0.05 gigatons of CO_2, or about 50 million metric tons. So the actual math involves a 50-million-ton blast versus more than 3,000 billion metric tons: you would need more than *60,000* Pinatubo eruptions (or 350,000 of President Reagan's preferred eruption, Mount St. Helens) to match a hundred years of human emissions.

If that isn't enough wrongness for you, it actually gets worse. When it comes to Huckabee's topic, global warming, the fact is that volcanoes actually have an opposite effect on the global climate system. The major result of massive eruptions like Pinatubo is to inject large amounts of sulfates high into the stratosphere—yes, just as Reagan spoke of, though he certainly got the numbers wrong. Those particles reflect sunlight back into space, and actually *cool* the Earth; the 1991 Pinatubo eruption resulted in a about a half-degree cooling effect over the following year. Huckabee brought up volcanoes to try to argue that human contributions to global warming are trivial compared to that of the natural world, when the exact opposite is true: humans are warming up the planet *in spite* of the world fighting back and trying to cool it down!

This STRAIGHT-UP FABRICATION, unfortunately, represented just another in a long line of climate falsehoods told by the TOADS. And even this bit of nonsense, off by a factor of sixty thousand or more, can be tough to spot; during the interview, Couric seemed to lack any data to refute Huckabee, and she simply let the enormous mistake stand. It's hard to blame her, given the lack of a specific source and the vast number of ways climate deniers have found to question the consensus.

———

THE VERY LAST EXAMPLE we'll explore in this book is another STRAIGHT-UP FABRICATION, one that has reared its head several times since the anti-vaccination movement sprang up, and is perhaps more damaging than virtually anything else a politician could say. It is quite literally a matter of life and death.

The original MMR vaccine scare centered around the idea that the immunization could be the cause of a child's autism spectrum disorder diagnosis. Even after more than a decade of countering evidence, the idea that vaccines could cause neurological or behavioral problems remained. In 2011, then-Congresswoman Michele Bachmann related an anecdote about the vaccine for human papillomavirus:

> The problem is, it comes with some very significant consequences. There's a woman who came up crying to me tonight after the debate. She said her daughter was given that vaccine. She told me her daughter suffered mental retardation as a result of that vaccine. There are very dangerous consequences.[13]

Though we know nothing about the specific person Bachmann cited, her goal was to scare people away from a lifesaving vaccine. And her reasoning, that Gardasil and other HPV vaccines can cause "mental retardation," is utterly lacking in scientific merit.

Though many were quick to debunk this mythical connection at the time, the idea would not die. In 2015, Senator Rand

Paul—an actual doctor, with an MD—repeated a very similar claim in an interview:

> I have heard of many tragic cases of walking, talking, normal children who wound up with profound mental disorders after vaccines.[14]

In a debate in September 2015, Donald Trump repeated much the same idea, using Bachmann's technique of telling an unverifiable story:

> Just the other day, two years old, two-and-a-half years old, a child, a beautiful child went to have the vaccine, and came back, and a week later got a tremendous fever. Got very, very sick, now is autistic.[15]

These are all just slightly altered versions of the old vaccine dog whistle—that the MMR causes autism—and they are just as untrue.

Neither the MMR nor the HPV vaccine has been shown to cause any sort of "profound mental disorder" or "mental retardation"; in fact, they have been proved to be remarkably safe. As with any drug or medication, there can be adverse effects, but with the vaccines available today, these are almost universally quite mild. There are reports of more severe reactions or events, but usually they are so rare that a cause-and-effect relationship cannot be determined.

This is not to say, though, that the scientific community has not taken the possibility of dangerous vaccines seriously.

It has, and has found that our immunizations are, in general, remarkably safe. A 2011 report by the Institute of Medicine summarized the safety of all the vaccines that the CDC currently recommends; the report specifically rejected the possibility of a connection between the MMR vaccine and autism.[16] It also found no evidence of a causal relationship between the MMR or DTaP (which protects against diphtheria, tetanus, and pertussis, also known as whooping cough) vaccines and type 1 diabetes, and between the flu vaccine and Bell's palsy, as well as worsening of asthma. The IOM found there *was* in fact evidence of such a relationship for a few adverse effects, including very rare cases of anaphylaxis (a severe allergic reaction) and the HPV vaccine, and other less dangerous problems. For the vast bulk of reported adverse effects, though, we lack anywhere near enough evidence to make a determination of causality. Here's how the IOM report concluded:

> Vaccines offer the promise of protection against a variety of infectious diseases. Despite much media attention and strong opinions from many quarters, vaccines remain one of the greatest tools in the public health arsenal. Certainly, some vaccines result in adverse effects that must be acknowledged. But the latest evidence shows that few adverse effects are caused by the vaccines reviewed in this report.

The weight of evidence is clear: the repeated claims that various vaccines have caused "mental retardation" or "profound mental disorders" have absolutely no basis in science.

Senator Paul walked back his comments in dramatic fashion in the ensuing days, even appearing in a *New York Times* photo as he received a hepatitis A vaccine booster.[17] But his STRAIGHT-UP FABRICATION didn't stop at the first blunder: that same day, he appeared on a radio show and dug his hole a bit deeper, proving that one of the only doctors in the Senate wasn't exactly on top of the current medical research:

> I was annoyed when my kids were born that they wanted them to take hepatitis B in the neonatal nursery, and it's like, that's a sexually transmitted disease, or a blood-borne disease, and I didn't like them getting 10 vaccines at once, so I actually delayed my kids' vaccines and had them staggered over time.[18]

First of all, regarding hepatitis B, the vaccine protecting against this disease is indeed administered in the neonatal nursery. There is a very good reason for this: hep B can be passed from mother to infant at birth, and the vaccine is extremely effective at preventing that transmission. Thus, it is the only vaccine commonly given at such a young age. Since the CDC began recommending the hepatitis B vaccine for all children in 1991, new infections among children and adolescents have dropped by an astonishing 95 percent or more.[19]

But what should we make of Paul's strategy to stagger his kids' vaccines? Paul sounded so reasonable on that count; wouldn't it make sense to try to space out all these substances being injected into a child's body? There are a lot of vaccines! Here's an "easy-to-read" version of the CDC's immunization schedule for birth through six years of age, highlighting just how many vaccines children get at about the same time:

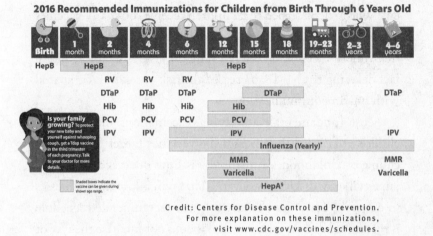

2016 Recommended Immunizations for Children from Birth Through 6 Years Old

Credit: Centers for Disease Control and Prevention. For more explanation on these immunizations, visit www.cdc.gov/vaccines/schedules.

As sensible as staggering vaccines may sound, it has no scientific basis whatsoever. For example, one 2013 study published in the *Journal of Pediatrics* examined whether exposure to more of the antigens found in vaccines correlated with autism spectrum disorders; they tested this in 256 children with such disorders and 752 control children with no such diagnosis.[20] Their results were remarkable: for every 25-unit increase in total antigen exposure, the risk of autism changed by essentially *zero*. (To be precise, it changed by a factor of 0.999, where 1.0 would mean no effect at all.)

Other studies have looked beyond autism to other broader "neuropsychological outcomes." One study in *Pediatrics* in 2010 found that timely vaccination—that is, vaccines administered within thirty days of the recommended age—was not associ-

ated with *any* negative outcomes at all out to ten years later.[21]
In fact, getting vaccinated on time actually improved some
neuropsychological outcomes. "These data may reassure parents who are concerned that children receive too many vaccines
too soon," the researchers concluded. Yet another study found
that timely vaccination did not raise the risk of seizures, and in
fact delaying the MMR increased the risk of seizures compared
with on-time immunization.[22]

Most fundamentally, delaying vaccinations means that a
child is at risk for the diseases the vaccines target for a bigger
chunk of childhood. "It's stupid," is how one vaccine expert,
James Cherry of UCLA, put it. "That will allow these illnesses
to occur."[23] This is not guesswork; the recommended schedule
is recommended for a reason, and it does not increase the risk of
problems. Rand Paul was completely, utterly wrong.

In spite of what the most vehement among the anti-vax
crowd might say, there is no grand conspiracy to pump "tox-
ins" into the world's children through vaccination programs.
Vaccines are one of science's greatest achievements and have
resulted in many, many thousands of lives saved over more than
a century; the CDC estimates that they saved 732,000 children's
lives *during just a twenty-year period*.[24] The science is clear: vacci-
nate your kids, and do it on time.

More generally, look out for the STRAIGHT-UP FABRICATION,
as hard as it can be to spot. If something sounds ridiculous—
"legitimate rape," or prison-based sexual orientation changes—
unleash your inner skeptic. And for the less ridiculous, more
reasonable-sounding claims on issues like vaccines, the only
antidote is to look for reputable sources and do your homework.
The politicians are hoping you won't; try to disappoint them.

The Conspicuous Silence

THE FIRST REPORT FROM THE CENTERS FOR DISEASE CONTROL and Prevention on what would eventually become known as acquired immune deficiency syndrome, or AIDS, was released in June of 1981.[1] Over the next couple of years, reports proliferated, research began in earnest, and the media began reporting on a new epidemic. The leader of the free world, meanwhile, said nothing.

Ronald Reagan famously did not publicly discuss AIDS until a press conference in 1985,[2] with a more comprehensive speech on the topic coming only in April 1987, as his presidency was winding down.[3] To many, Reagan's lack of response to the crisis represented an implicit condemnation of homosexuality, since at the time the disease was considered a "plague" targeting mostly gay people. The president didn't CHERRY-PICK data, or claim the science was unsettled, or make any errors at all regarding a complicated and rapidly evolving scientific issue: he simply didn't mention it.

This bonus error is the CONSPICUOUS SILENCE, and it is just as big a scourge as the other errors in this book. Failing to address issues of importance spreads the idea that those issues don't matter. Why should the public care about something if our elected officials, up to and including the president, don't seem to? Talking about science can normalize it, though of course the not-scientists doing the talking should lean heavily on the actual scientists, who know what they're talking about. It is a simple two-step process that many in politics seem to ignore: learn what is known, then talk about it.

In Reagan's case, critics say his silence on AIDS (HIV, the virus that causes AIDS, was not named until May 1986) greatly hindered research into its causes, potential treatments, and prevention efforts.[4] A year after the president's first public mention of the issue, a surgeon general's report noted that 1.5 million people in the United States were likely infected.[5] In that first press conference, Reagan defended his administration's funding for AIDS research, which included a request for $126 million for the following year. A top government scientist, though, reportedly called the Reagan administration funding "not nearly enough at this stage to go forward and really attack the problem."[6] In fact, Congress voted to boost the presidential budget request for three consecutive years, showing that at least some elected officials seemed to have a better grasp of the problem.

The modern era of HIV management began in 1995, with the advent of HAART—highly active anti-retroviral therapy. The drugs used in HAART, along with more recently developed drugs, can often keep virus levels down to near-background levels, extending the lives of individuals for many years longer than was previously possible. By the end of 1995, more than 319,000 deaths[7] from AIDS had been reported in the United States. It is, of course, impossible to say whether HAART could have come along sooner, or whether public health prevention methods could have been amplified, with earlier public input from the president. But talking about it— carefully, with expert help—would not have made the situation worse.

Today, many are pushing for just that sort of talk therapy for science and politics. One way to usher science into the limelight is a science-only televised debate between political candidates—an idea created in 2008 and championed ever since by a small group of journalists, science professors, filmmakers, and others.[8] During the 2012 general election, though an actual live science debate did not materialize, the group did manage to get President Obama and GOP nominee Mitt Romney on the record answering a series of science-based questions in print.[9] It served to put each of the candidates' science-related policy positions in the spotlight, though obviously it did not have the audience that a televised debate would have.

Some of the answers actually did contain some errors, including some of the type discussed in this book. On a question regarding climate change, Romney even began by saying: "I am not a scientist myself." He did go on to say that his "best assessment of the data is that the world is getting warmer, that human activity contributes to that warming, and that policymakers should therefore consider the risk of negative consequences," meaning he at least was willing to listen to scientists. He then questioned whether there was a "scientific consensus" on the human contribution (an example of the CERTAIN UNCER-

TAINTY we saw from Jeb Bush as well), the extent of warming, and the "severity of risk," and still opposed most actions to reduce emissions.

I asked the founders and directors of Science Debate, the nonprofit group behind this effort, why they think a science-only debate would help improve scientific discourse. Lawrence Krauss, a professor of theoretical physics at Arizona State University and author of a number of popular books on science, said that "there is no opportunity cost right now associated with ignoring" important scientific topics. "The marginalization of science makes it perfectly acceptable for candidates to be ignorant about the issues, or to lie about them, or obfuscate them."[10]

That challenge has only grown in recent years. "Anti-science is worse now in political circles than at any time since the 1910s and 1920s," Shawn Lawrence Otto, an author and screenwriter who cofounded Science Debate, told me. "Since complex sciences influence every aspect of life on the planet, we've got to find a smarter way of incorporating it into our policy dialogue," Otto said.[11]

Normalizing science and discussion of science would likely do wonders for reducing the potential for political missteps. Many of the errors in this book could be avoided simply by letting the scientific consensus act as a talking point; of course, the scientific consensus has to be largely understood in order for that to work. And as long as there is money in politics and votes are at stake, politicians with an agenda that runs counter to the best available science will attempt to undermine it. If we, the general public, have more exposure to the issues in question, though, those attempts may not slip by us so easily.

IN JANUARY 2015, President Obama gave his State of the Union address to both houses of Congress, and to millions of viewers at home. In a section of his speech dedicated to urging action on climate change, he decided to absolutely murder the talking point that serves as the title of this book:

> I've heard some folks try to dodge the evidence by say-ing they're not scientists; that we don't have enough information to act. *Well, I'm not a scientist, either. But you know what, I know a lot of really good scientists* at NASA, and at NOAA, and at our major universities. And the best scientists in the world are all telling us that our activities are changing the climate, and if we don't act forcefully, we'll continue to see rising oceans, longer, hotter heat waves, dangerous droughts and floods, and massive dis-ruptions that can trigger greater migration and conflict and hunger around the globe.[12]

To any politician paying attention—including some users of the line sitting in the room during the president's speech—the "I'm not a scientist" dodge was officially dead. It likely would have died out on its own eventually, given that even those on the team using it considered it "the dumbest talking point in the history of mankind."[13] But the president surely hastened the demise and, in the process, made a convincing case for simply *listening* to people who really are experts.

Politics being what it is, however, our leaders are far more likely to listen to *us*, the voting public. That's why it is increasingly crucial that we all join in the effort to call out the mistakes politicians make on science. This playbook of rhetorical and logical errors has given you some ammunition in the fight against scientific ignorance and misinformation. Use it! If you hear your senator or congressperson spouting off about "unsettled science," or some ultraspecific subset of climate data, or fearmongering about diseases and immigration, call them out. Did you catch a CREDIT SNATCH, or a RIDICULE AND DISMISS? Take to Twitter and Facebook and anywhere else you can reach an audience; get #NotAScientist trending, hold the politicians accountable for their errors, and in the process help *actual* scientists move their work forward in ways that truly help the world.

Barring some unforeseen massive overhaul in political circles, politicians will likely continue to make the errors outlined in this book. The issues themselves may change—as we get a handle on climate change, as new diseases arise and spread, or as topics such as nanotechnology and antibiotic resistance become more central to everyday life—but the missteps, fallacies, and techniques used are timeless. As they have told you many times by now, your elected officials are not scientists. Keep that in mind as they try to sneak their not-science past you.

Notes

....

INTRODUCTION

1. D. E. Kneeland, "Carter Softens His Criticism of Foe; Reagan Defends Record on Ecology: Teamsters Back Republican," *New York Times*, October 10, 1980 (italics mine).
2. S. J. Smith, J. van Aardenne, Z. Klimont, et al., "Anthropogenic Sulfur Dioxide Emissions: 1850–2005," *Atmospheric Chemistry and Physics* 11 (2011): 1101–16, http://www.atmos-chem-phys .net/11/1101/2011/acp-11-1101-2011.pdf.
3. T. M. Gerlach and K. A. McGee, "Total Sulfur Dioxide Emissions and Pre-eruption Vapor-Saturated Magma at Mount St. Helens, 1980–88," *Geophysical Research Letters* 21 (1994): 2833–36, http:// onlinelibrary.wiley.com/doi/10.1029/94GL02761/abstract.
4. Environmental Protection Agency, "Sulfur Dioxide," http:// www3.epa.gov/airquality/sulfurdioxide, accessed April 2, 2016.
5. B. Reinhard, "Marco Rubio Accused of Switching Stance on Global Warming," *Miami Herald*, December 10, 2009.
6. M. Caputo, "Rick Scott Won't Say If He Thinks Man-Made Climate-Change Is Real, Significant," *Naked Politics* (blog), *Miami Herald*, May 27, 2014, http://miamiherald.typepad.com/naked politics/2014/05/rick-scott-wont-say-if-he-thinks-man-made -climate-change-is-real-significant.html.
7. A. Kaczynski, "John Boehner: I'm Not Qualified to Debate Science over Climate Change," *BuzzFeed News*, May 29, 2014, http://

www.buzzfeed.com/andrewkaczynski/john-boehner-im-not
-qualified-to-debate-science-over-climate#.hqrlJl5G1N.

8. S. Wartman, "McConnell Talks Brent Spence, Heroin, Ebola," *Cincinnati Herald*, October 2, 2014, http://www.cincinnati.com/ story/news/politics/2014/10/02/mcconnell-taking-ebola-threat -seriously/16586749.

9. C. Richert, "Minnesota Republicans Change Their Tone on Climate Change," *MPR News*, June 23, 2015, http://www.mprnews .org/story/2015/06/23/climate-change-republicans.

10. C. Mooney, *The Republican War on Science* (Basic Books, 2006).

11. L. McGaughy, "Jindal Full Remarks to Republican National Committee Meeting," *New Orleans Times Picayune*, January 24, 2013, http://www.nola.com/politics/index.ssf/2013/01/jindal_full _remarks_to_republi.html.

12. N. Geiling, "Lindsey Graham Challenges Republicans: 'Tell Me Why' You Deny Climate Science," *ThinkProgress* (blog), October 12, 2015, http://thinkprogress.org/climate/2015/10/12/3711587/lind sey-graham-no-labels-climate-change.

CHAPTER 1: THE OVERSIMPLIFICATION

1. "House Session," May 13, 2015, C-SPAN, http://www.c-span.org/ video/?325947-2/us-house-debate-20week-abortion-ban.

2. "Fetal Awareness: Review of Research and Recommendations for Practice," Royal College of Obstetricians and Gynaecologists, March 2010, https://www.rcog.org.uk/globalassets/documents/guide lines/rcogfetalawarenesswpr0610.pdf.

3. L. M. Hollier, "Texas-ACOG Statement Opposes Texas Fetal Pain Legislation," April 10, 2013, http://www.acog.org/~/media/ Departments/State%20Legislative%20Activities/20130410CookLtr .pdf?dmc=1&ts=20130823T2103524859.

4. S. J. Lee, H. J. P. Ralston, E. A. Drey, et al., "Fetal Pain: A Systematic Multidisciplinary Review of the Evidence," *Journal of the American Medical Association* 294 (2005): 947-54, http://jama .jamanetwork.com/article.aspx?articleid=201429#.

5. B. Merker, "Consciousness without a Cerebral Cortex: A Challenge for Neuroscience and Medicine," *Behavioral and Brain Sciences*

30 (2007): 63–81, http://journals.cambridge.org/action/displayAbs tract?fromPage=online&aid=1007572.

6. P. Belluck, "Complex Science at Issue in Politics of Fetal Pain," *New York Times*, September 16, 2013, http://www.nytimes.com/ 2013/09/17/health/complex-science-at-issue-in-politics-of-fetal -pain.html?pagewanted=2&_r=1&hp.

7. Lee et al., "Fetal Pain."

8. Cole Avery (in Representative Ralph Abraham's office), e-mail to the author, quoted in D. Levitan, "Does a Fetus Feel Pain at 20 Weeks?" FactCheck.org, May 18, 2015, http://www.factcheck .org/2015/05/does-a-fetus-feel-pain-at-20-weeks.

9. H. Hewitt, "Governor Christie Bear Hugs the Third Rail," *HughHewitt* (blog), April 14, 2015, http://www.hughhewitt.com/ governor-christie-bear-hugs-the-third-rail.

10. Susan Weiss (NIDA), phone interview with the author, quoted in D. Levitan, "Is Marijuana Really a 'Gateway Drug'?" April 23, 2015, http://www.factcheck.org/2015/04/is-marijuana-really-a -gateway-drug.

11. J. E. Joy, S. J. Watson Jr., and J. A. Benson, eds., *Marijuana and Medicine: Assessing the Science Base* (National Academies Press, 1999), 100–101.

12. W. Moser, "Heat and Crime: It's Not Just You Feeling It," *Chicago*, March 15, 2012, http://www.chicagomag.com/Chicago-Maga zine/The-312/March-2012/Heat-and-Crime-Its-Not-Just-You -Feeling-It.

13. M. Ellgren, S. M. Spano, and Y. L. Hurd, "Adolescent Cannabis Exposure Alters Opiate Intake and Opioid Limbic Neuronal Populations in Adult Rats," *Neuropsychopharmacology* 32 (2007): 607–15, http://www.nature.com/npp/journal/v32/n3/full/1301127a .html.

14. S. Stopponi, L. Soverchia, M. Ubaldi, et al., "Chronic THC during Adolescence Increases the Vulnerability to Stress-Induced Relapse to Heroin Seeking in Adult Rats," *European Neuropsychopharmacology* 24 (2014): 1037–45, http://www.europeanneuropsych opharmacology.com/article/S0924-977X(13)00357-X/abstract.

15. M. Pistis, S. Perra, G. Pillolla, et al., "Adolescent Exposure to Cannabinoids Induces Long-Lasting Changes in the Response to

Drugs of Abuse of Rat Midbrain Dopamine Neurons," *Biological Psychiatry* 56 (2004): 86–94, http://www.biologicalpsychiatryjournal .com/article/S0006-3223(04)00530-X/abstract.

16. M. T. Lynskey, A. C. Heath, K. K. Buchholz, et al., "Escalation of Drug Use in Early-Onset Cannabis Users vs Co-twin Controls," *Journal of the American Medical Association* 289 (2003): 427–33, http:// jama.jamanetwork.com/article.aspx?articleid=195839.

17. Weiss interview, April 23, 2015.

18. H. H. Cleveland and R. P. Wiebe, "Understanding the Association between Adolescent Marijuana Use and Later Serious Drug Use: Gateway Effect or Developmental Trajectory?" *Development and Psychopathology* 20 (2008): 615–32, http://journals.cambridge.org/ action/displayAbstract?fromPage=online&aid=1842736&fileId=S0 954579408000308.

19. David Fergusson (University of Otago), e-mail to the author, quoted in Levitan, "Is Marijuana Really a Gateway Drug?"

20. R. J. MacCoun, "What Can We Learn from the Dutch Cannabis Coffeeshop Experience?" Rand Drug Policy Research Center, July 2010, http://www.rand.org/content/dam/rand/pubs/work ing_papers/2010/RAND_WR768.pdf.

21. White House, Office of the Press Secretary, "Remarks by the President in State of the Union Address," January 20, 2015, https://www .whitehouse.gov/the-press-office/2015/01/20/remarks-president -state-union-address-january-20-2015.

22. NOAA, "Global Analysis—December 2014," http://www.ncdc .noaa.gov/sotc/global/201412, accessed April 2, 2016.

23. G. A. Schmidt and T. R. Karl, "Annual Global Analysis for 2014," NOAA/NASA, January 2015, http://www.ncdc.noaa.gov/sotc/ briefings/201501.pdf.

24. A. C. Revkin, "How 'Warmest Ever' Headlines and Debates Can Obscure What Matters about Climate Change," *Dot Earth* (blog), *New York Times*, January 21, 2015, http://dotearth.blogs.nytimes .com/2015/01/21/how-warmest-ever-headlines-and-debates-can -obscure-what-matters-about-climate-change/?_r=1.

25. D. Rose, "NASA Climate Scientists: We Said 2014 Was the Warmest Year on Record . . . but We're Only 38% Sure We Were Right," *Daily Mail*, January 17, 2015, http://www.dailymail.co.uk/news/

article-2915061/Nasa-climate-scientists-said-2014-warmest-year
-record-38-sure-right.html.

26. White House, Office of the Press Secretary, "Remarks by the President," January 20, 2015.

27. "NASA, NOAA Analyses Reveal Record-Shattering Global Warm Temperatures in 2015," NASA, January 20, 2016, http://www.giss .nasa.gov/research/news/20160120.

CHAPTER 2: THE CHERRY-PICK

1. US Senate Committee on Environment & Public Works, "Inhofe in Copenhagen: 'It Has Failed . . . It's Déjà Vu All Over Again'" (press release, December 17, 2009), http://www.epw.senate.gov/ public/index.cfm/press-releases-all?ID=9cac1e35-802a-23ad-4540 -3e4706eab1bd&Region_id=&Issue_id=.

2. "Senator James Inhofe Throws Snowball on Senate Floor," C-SPAN, February 26, 2015, http://www.c-span.org/video/?c4529383/senator -james-inhofe-throws-snowball-senate-floor.

3. P. Colford, "An Addition to AP Stylebook Entry on Global Warming," *The Definitive Source* (blog), Associated Press, September 22, 2015, https://blog.ap.org/announcements/an-addition-to -ap-stylebook-entry-on-global-warming.

4. G. A. Meehl, C. Tebaldi, G. Walton, et al., "Relative Increase of Record High Maximum Temperatures Compared to Record Low Minimum Temperatures in the U.S.," *Geophysical Research Letters* 36 (2009), http://onlinelibrary.wiley.com/doi/10.1029/2009GL040736/ abstract.

5. Climate Central, "Record Highs vs. Record Lows," August 19, 2015, http://www.climatecentral.org/gallery/graphics/record-highs -vs-record-lows.

6. J. Root and T. Wiseman, "Video: One-on-One Interview with Ted Cruz," *Texas Tribune*, March 24, 2015, https://www.texastribune .org/2015/03/24/livestream-one-on-one-interview-with-ted-cruz.

7. T. R. Karl, A. Arguez, B. Huang, et al., "Possible Artifacts of Data Biases in the Recent Global Surface Warming Hiatus," *Science* 348 (2015): 1469–72, http://www.sciencemag.org/content/348/6242/1469 .abstract.

8. Associated Press, "New Fed Data Shows No Stopping or Slowing of Global Warming," *New York Times*, June 4, 2015, http://www .nytimes.com/aponline/2015/06/04/science/ap-us-sci-warming -hiatus.html?_r=0.

9. J. C. Fyfe, G. A. Meehl, M. H. England, et al., "Making Sense of the Early-2000s Warming Slowdown," *Nature Climate Change* 6 (2016): 224–28, http://www.nature.com/nclimate/journal/v6/n3/ full/nclimate2938.html.

10. J. Christy and R. Spencer, "Washington Roundtable on Science & Public Policy: Satellite Temperature Data," George C. Marshall Institute, April 17, 2006, http://marshall.wpengine.com/ wp-content/uploads/2013/08/Christy-and-Spencer-Satellite -Temperature-Data.pdf; George C. Marshall Institute, "Board Members," http://marshall.org/board-members, accessed April 13, 2016.

11. J. R. Christy, "My Nobel Moment," *Wall Street Journal*, November 1, 2007, http://www.wsj.com/articles/SB119387567378878423.

12. R. Pierrehumbert, "How to Cook a Graph in Three Easy Lessons," *RealClimate* (blog), May 21, 2008, http://www.realclimate .org/index.php/archives/2008/05/how-to-cook-a-graph-in-three -easy-lessons.

13. S. Po-Chedley, T. J. Thorsen, and Q. Fu, "Removing Diurnal Cycle Contamination in Satellite-Derived Tropospheric Temperatures: Understanding Tropical Tropospheric Trend Discrepancies," *Journal of Climate* 28 (2015): 2274–90, http://journals .ametsoc.org/doi/abs/10.1175/JCLI-D-13-00767.1.

14. C. Mooney, "Ted Cruz Says Satellite Data Show the Globe Isn't Warming. This Satellite Scientist Feels Otherwise," *Washington Post*, March 24, 2015, https://www.washingtonpost.com/news/ energy-environment/wp/2015/03/24/ted-cruz-says-satellite- data-show-the-globe-isnt-warming-this-satellite-scientist-feels -otherwise.

15. P. Plait, "When Denial Attacks: Ted Cruz vs. Reality," *Bad Astronomy* (blog), *Slate*, October 8, 2015, http://www.slate.com/ blogs/bad_astronomy/2015/10/08/ted_cruz_grills_sierra_club_ president_with_a_firehose_of_denial.html.

16. "Sarah Palin on State of the Union: Full Interview," CNN, Sep-

tember 6, 2015, http://www.cnn.com/videos/politics/2015/09/06/sotu-tapper-sarah-palin-full-interview.cnn.

17. S. Huse, "The Retreat of Exit Glacier," National Park Service, http://www.nps.gov/kefj/learn/nature/upload/The%20Retreat%20of%20Exit%20Glacier.pdf, accessed April 13, 2016.

18. Barack Obama, "My Visit to Alaska: The Signpost for Climate Change," Medium, September 2, 2015, https://medium.com/@PresidentObama/my-visit-to-alaska-the-signpost-for-climate-change-bc75bd96aa12#.19gs83u31.

19. "The Advance of Hubbard Glacier," *Earth Observatory* (blog), NASA, May 20, 2015, http://earthobservatory.nasa.gov/IOTD/view.php?id=85900.

20. L. A. Stearns, G. S. Hamilton, C. J. van der Veen, et al., "Glaciological and Marine Geological Controls on Terminus Dynamics of Hubbard Glacier, Southeast Alaska," *Journal of Geophysical Research*, 120 (2015): 1065–81, http://onlinelibrary.wiley.com/doi/10.1002/2014JF003341/full.

21. Empire State Realty Trust, "Empire State Building Fact Sheet," http://www.esbnyc.com/sites/default/files/esb_fact_sheet_4_9_14_4.pdf, accessed April 13, 2016.

22. C. Mooney, "To Truly Grasp What We're Doing to the Planet, You Need to Understand This Gigantic Measurement," *Washington Post*, July 1, 2015, http://www.washingtonpost.com/news/energy-environment/wp/2015/07/01/meet-the-gigaton-the-huge-unit-that-scientists-use-to-track-planetary-change.

23. World Glacier Monitoring Service, "Latest Glacier Mass Balance Data," http://wgms.ch/latest-glacier-mass-balance-data, accessed April 13, 2016.

24. National Snow and Ice Data Center, "The Life of a Glacier," https://nsidc.org/cryosphere/glaciers/life-glacier.html, accessed April 13, 2016.

25. C. Mooney, "Greenland Has Lost a Staggering Amount of Ice—and It's Only Getting Worse," *Washington Post*, December 16, 2015, https://www.washingtonpost.com/news/energy-environment/wp/2015/12/16/greenland-has-lost-a-staggering-amount-of-ice-and-its-only-getting-worse.

CHAPTER 3: THE BUTTER-UP AND UNDERCUT

1. US Senate Subcommittee on Space, Science, and Competitiveness, "Examining the President's Fiscal Year 2016 Budget Request for the National Aeronautics and Space Administration," March 12, 2015, http://www.commerce.senate.gov/public/index.cfm?p=Hearings&ContentRecord_id=4ccea8c1-af33-4439-ba02-075de42c9946&ContentType_id=14f995b9-dfa5-407a-9d35-56cc7152a7ed&Group_id=a06730c4-d875-4fde-97db-9e2be611840e.

2. S. Motel, "NASA Popularity Still Sky-High," Pew Research Center, February 3, 2015, http://www.pewresearch.org/fact-tank/2015/02/03/nasa-popularity-still-sky-high.

3. US Senate Subcommittee on Space, Science, and Competitiveness, "Examining the President's Fiscal Year 2016 Budget Request."

4. National Aeronautics and Space Administration, "National Aeronautics and Space Act of 1958 (Unamended)," http://history.nasa.gov/spaceact.html, accessed April 13, 2016.

5. National Aeronautics and Space Administration, "Budget Estimates: Fiscal Year 1965, Volume I: Summary Data," 1964, http://www.hq.nasa.gov/office/hqlibrary/documents/Budgets/03490366_1965_1.pdf.

6. National Aeronautics and Space Administration, "Budget Estimates: Fiscal Year 1975, Volume I: Agency Summary, Research and Development," 1974, http://www.hq.nasa.gov/office/hqlibrary/documents/Budgets/03490366_1975_1.pdf.

7. Global Climate Change: Vital Signs of the Planet, "Taking a Global Perspective on Earth's Climate," NASA, http://climate.nasa.gov/nasa_role, accessed April 13, 2016.

8. National Aeronautics and Space Administration, "Budget Estimates: Fiscal Year 1990, Volume I: Agency Summary; Research and Development; Space Flight, Control and Data Communications," [1989], http://www.hq.nasa.gov/office/hqlibrary/documents/Budgets/03490366_1990_pt_1.pdf.

9. US Senate Subcommittee on Space, Science, and Competitiveness, "Examining the President's Fiscal Year 2016 Budget Request."

10. C. Rosenzweig, R. M. Horton, D. A. Bader, et al., "Enhancing

Climate Resilience at NASA Centers: A Collaboration between Science and Stewardship," *Bulletin of the American Meteorological Society* 95 (2014): 1351–63, http://journals.ametsoc.org/doi/full/10.1175/BAMS-D-12-00169.1.

11. C. S. Watson, N. J. White, J. A. Church, et al., "Unabated Global Mean Sea-Level Rise over the Satellite Altimeter Era," *Nature Climate Change* 5 (2015): 565–69, http://www.nature.com/nclimate/journal/v5/n6/full/nclimate2635.html.

12. Rosenzweig et al., "Enhancing Climate Resilience at NASA Centers."

13. US Senate Subcommittee on Space, Science, and Competitiveness, "Examining the President's Fiscal Year 2016 Budget Request."

14. C. W. McEntee (executive director/CEO, American Geophysical Union) to Senator Ted Cruz, March 13, 2015, http://sciencepolicy.agu.org/files/2013/07/AGU-letter-to-Sen-Cruz-March-2015.pdf.

15. National Research Council, *Earth Science and Applications from Space: National Imperatives for the Next Decade and Beyond* (National Academies Press, 2007).

16. World Health Organization, "FAQs: H5N1 Influenza," http://www.who.int/influenza/human_animal_interface/avian_influenza/h5n1_research/faqs/en, accessed April 13, 2016.

17. World Health Organization, "H5N1 Avian Influenza: Timeline of Major Events," January 25, 2012, http://www.who.int/influenza/human_animal_interface/H5N1_avian_influenza_update.pdf.

18. "Transcript of Bush Speech on Pandemic Flu Strategy," November 1, 2005, CNN, http://www.cnn.com/2005/HEALTH/conditions/11/01/bush.transcript.

19. National Institutes of Health, "History of Congressional Appropriations, Fiscal Years 2000–2013," https://officeofbudget.od.nih.gov/pdfs/FY15/Approp%20%20History%20by%20IC%20through%20FY%202013.pdf, accessed April 13, 2016.

20. D. Nather, "It's Official: The NIH Budget Is Getting an Extra $2 Billion," STAT, December 18, 2015, http://www.statnews.com/2015/12/18/nih-increase-congress-vote.

21. National Institutes of Health, "Research Project Success Rates by Type and Activity for 2014," http://report.nih.gov/success_rates/Success_ByActivity.cfm, accessed April 13, 2016.

22. "H.R.3043—Departments of Labor, Health and Human Services, and Education, and Related Agencies Appropriations Act, 2008," Congress.gov, https://www.congress.gov/bill/110th-congress/house-bill/3043, accessed April 13, 2016.

23. White House, Office of the Press Secretary, "President Bush Visits Indiana, Discusses Budget," November 13, 2007, https://web.archive.org/web/20071114111403/http://www.whitehouse.gov/news/releases/2007/11/20071113-7.html.

24. J. Hicks, "NIH Director: Ebola Vaccine Could Be Ready by Now If Not for Budget Austerity," *Washington Post*, October 13, 2014, https://www.washingtonpost.com/news/federal-eye/wp/2014/10/13/nih-director-ebola-vaccine-could-be-ready-if-not-for-budget-austerity.

CHAPTER 4: THE DEMONIZER

1. M. S. Majumder, E. L. Cohn, S. R. Mekaru, et al., "Substandard Vaccination Compliance and the 2015 Measles Outbreak," *JAMA Pediatrics* 169 (2015): 494–95, http://archpedi.jamanetwork.com/article.aspx?articleid=2203906.

2. Matt Murphy Show, "Mo Brooks Says Immigrants Causing Measles Outbreak," February 3, 2015, available through SoundCloud at ThinkProgress.org, http://thinkprogress.org/immigration/2015/02/03/3618647/mo-brooks-measles-immigrants.

3. Centers for Disease Control and Prevention, "Frequently Asked Questions about Measles in the U.S.," http://www.cdc.gov/measles/about/faqs.html#measles-elimination, accessed April 13, 2016.

4. Centers for Disease Control and Prevention, "Enterovirus D68," http://www.cdc.gov/non-polio-enterovirus/about/ev-d68.html, accessed April 13, 2016.

5. US Department of Health & Human Services, Administration for Children & Families, Office of Refugee Settlement, "Unaccompanied Children Frequently Asked Questions," http://www.acf.hhs.gov/unaccompanied-children-frequently-asked-questions, accessed April 13, 2016.

6. J. Garcia, V. Espejo, M. Nelson, et al., "Human Rhinoviruses and

Enteroviruses in Influenza-Like Illness in Latin America," *Virology Journal* 10 (2013): 305, http://virologyj.biomedcentral.com/articles/10.1186/1743-422X-10-305.

7. A. Wakefield, S. H. Murch, A. Anthony, et al., "Ileal-Lymphoid-Nodular Hyperplasia, Non-specific Colitis, and Pervasive Developmental Disorder in Children," *Lancet* 351 (1998): 637–41, http://www.thelancet.com/pdfs/journals/lancet/PIIS0140-6736%2897%2911096-0.pdf. Retracted in 2010 by the editors, http://www.thelancet.com/journals/lancet/article/PIIS0140-6736(10)60175-4/abstract.

8. B. Deer, "How the Case against the MMR Vaccine Was Fixed," *British Medical Journal* 342 (2011): c5347, http://www.bmj.com/content/342/bmj.c5347.

9. For a full and thorough discussion of the anti-vaccination movement, read journalist Seth Mnookin's 2012 book *The Panic Virus: The True Story behind the Vaccine-Autism Controversy* (Simon & Schuster).

10. World Health Organization, "Immunization Surveillance, Assessment and Monitoring: Measles 1st Dose (MCV1) Immunization Coverage among 1-Year Olds, 1980–2014 (%): 2014," http://gamapserver.who.int/gho/interactive_charts/immunization/mcv/atlas.html, accessed April 13, 2016.

11. P. Esuivel and S. Poindexter, "Plunge in Kindergartners' Vaccination Rate Worries Health Officials," *Los Angeles Times*, September 2, 2014, http://www.latimes.com/local/education/la-me-school-vaccines-20140903-story.html.

12. C. Ingraham, "California's Epidemic of Vaccine Denial, Mapped," *Wonkblog, Washington Post*, January 27, 2015, https://www.washingtonpost.com/news/wonk/wp/2015/01/27/californias-epidemic-of-vaccine-denial-mapped.

13. California Department of Public Health, Immunization Branch, "2015–2016 Kindergarten Immunization Assessment," http://www.cdph.ca.gov/programs/immunize/Documents/2015-16_CA_KindergartenSummaryReport.pdf, accessed April 13, 2016.

14. E. Bradner, "Carson: No Exemptions on Immunizations," CNN, February 4, 2015, http://www.cnn.com/2015/02/03/politics/measles-vaccines-ben-carson-immunization.

15. K. Drew, "This Is What Trump's Border Wall Could Cost US," CNBC, October 9, 2015, http://www.cnbc.com/2015/10/09/this-is -what-trumps-border-wall-could-cost-us.html.

16. P. J. Buchanan, *State of Emergency: The Third World Invasion and Conquest of America* (St. Martin's Griffin, 2007), 29.

17. Centers for Disease Control and Prevention, "Malaria Facts: Malaria in the United States," http://www.cdc.gov/malaria/about/facts.html, accessed April 13, 2016.

18. Centers for Disease Control and Prevention, "Polio Elimination in the United States," http://www.cdc.gov/polio/us/index.html, accessed April 13, 2016.

19. K. V. Kowdley, C. C. Wang, S. Welch, et al., "Prevalence of Chronic Hepatitis B among Foreign-Born Persons Living in the United States by Country of Origin," *Hepatology* 56 (2012): 422–33, http://onlinelibrary.wiley.com/doi/10.1002/hep.24804/abstract.

20. G. Pearson, "Remain Calm: Kissing Bugs Are Not Invading the US," *Wired*, December 3, 2015, http://www.wired.com/2015/12/remain-calm-kissing-bugs-are-not-invading-the-us.

21. Centers for Disease Control and Prevention, "Incidence of Hansen's Disease—United States, 1994–2011," *Morbidity and Mortality Weekly Report* 63 (2014): 969–72, http://www.cdc.gov/mmwr/preview/mmwrhtml/mm6343a1.htm.

22. R. W. Truman, P. Singh, R. Sharma, et al., "Probable Zoonotic Leprosy in the Southern United States," *New England Journal of Medicine* 364 (2011): 1626–33, http://www.nejm.org/doi/full/10.1056/NEJMoa1010536.

23. "Senate Session," February 17, 1993, C-Span, http://www.c-span.org/video/?38148-1/senate-session.

24. T. Miller, "Obama Announces End of HIV Travel Ban," PBS NewsHour, October 30, 2009, http://www.pbs.org/newshour/updates/politics-july-dec09-travel_10-30.

25. J. N. Burns, R. Acuna-Soto, and D. W. Stahle, "Drought and Epidemic Typhus, Central Mexico, 1655–1918," *Emerging Infectious Diseases* 20 (2014), http://wwwnc.cdc.gov/eid/article/20/3/13-1366 _article.

26. Miguel A. Levario, interview by Christopher Rose on *15 Minute History*, "Episode 25: Mexican Migration to the US," September 4,

2013, http://15minutehistory.org/2013/09/04/mexican-migration-to
-the-us.

27. Immigration Act of 1917, 64th Congr. (1917), http://library.uwb
.edu/static/USimmigration/39%20stat%20874.pdf (italics mine).

28. H. Markel and A. M. Stern, "The Foreignness of Germs: The Persistent Association of Immigrants and Disease in American Society," *Milbank Quarterly* 80 (2002): 757–88, http://www.ncbi.nlm
.nih.gov/pmc/articles/PMC2690128.

CHAPTER 5: THE BLAME THE BLOGGER

1. Gary Palmer, interviewed on the Matt Murphy Show (no longer
accessible online). Originally transcribed by the author and printed
in the following story: D. Levitan, "Nothing False about Temperature Data," FactCheck.org, February 12, 2015, http://www
.factcheck.org/2015/02/nothing-false-about-temperature-data.

2. J. Delingpole, "Forget Climategate: This 'Global Warming' Scandal Is Much Bigger," Breitbart, January 30, 2015, http://www
.breitbart.com/london/2015/01/30/forget-climategate-this-global
-warming-scandal-is-much-bigger.

3. P. Homewood, "Massive Tampering with Temperatures in South
America," *Not a Lot of People Know That* (blog), January 20, 2015,
https://notalotofpeopleknowthat.wordpress.com/2015/01/20/
massive-tampering-with-temperatures-in-south-america.

4. C. Booker, "The Fiddling with Temperature Data Is the Biggest Science Scandal Ever," *Telegraph*, February 7, 2015, http://
www.telegraph.co.uk/news/earth/environment/globalwarming/
11395516/The-fiddling-with-temperature-data-is-the-biggest
-science-scandal-ever.html.

5. Berkeley Earth, "About Berkeley Earth," http://berkeleyearth.org/
about, accessed April 13, 2016.

6. R. A. Muller, "The Conversion of a Climate-Change Skeptic,"
New York Times, July 28, 2012, http://www.nytimes.com/2012/07/
30/opinion/the-conversion-of-a-climate-change-skeptic.html.

7. C. N. Williams, M. J. Menne, and P. W. Thorne, "Bench-
marking the Performance of Pairwise Homogenization of Surface Temperatures in the United States," *Journal of Geophysical*

Research 117 (2012): D05116, http://onlinelibrary.wiley.com/doi/10
.1029/2011JD016761/full.

8. M. J. Menne, C. N. Williams Jr., and M. A. Palecki, "On the
Reliability of the U.S. Surface Temperature Record," *Journal of
Geophysical Research* 115 (2010): D11108, http://onlinelibrary.wiley
.com/doi/10.1029/2009JD013094/abstract.

9. J. H. Lawrimore, M. J. Menne, B. E. Gleason, et al., "An Overview of the Global Historical Climatology Network Monthly
Mean Temperature Data Set, Version 3," *Journal of Geophysical
Research* 116 (2011): D19121, http://onlinelibrary.wiley.com/doi/
10.1029/2011JD016187/full.

10. Both of these searches also now call up a FactCheck.org story by
the author from February 2015 (Levitan, "Nothing False about
Temperature Data"), meaning there is some good information
among the bad.

11. "Real Time with Bill Maher: Rick Santorum—August 28, 2015,"
YouTube, August 28, 2015, https://www.youtube.com/watch?v=
CMcpz87EahU.

12. Global Climate Change: Vital Signs of the Planet, "Scientific
Consensus: Earth's Climate Is Warming," NASA, http://climate
.nasa.gov/scientific-consensus, accessed April 13, 2016.

13. B. Strengers, B. Verheggen, and K. Vringer, "Climate Science
Survey: Questions and Responses," PBL Netherlands Environmental Assessment Agency, April 10, 2015, http://www.pbl.nl/
sites/default/files/cms/publicaties/pbl-2015-climate-science-sur
vey-questions-and-responses_01731.pdf.

14. "The 97% Consensus of Climate Scientists Is Only 47%," *Fabius
Maximus* (blog), July 29, 2015, http://fabiusmaximus.com/2015/07/
29/new-study-undercuts-ipcc-keynote-finding-87796.

15. J. Root and T. Wiseman, "Video: One-on-One Interview with
Ted Cruz," *Texas Tribune*, March 24, 2015, https://www.texas
tribune.org/2015/03/24/livestream-one-on-one-interview-with
-ted-cruz.

16. P. Gwynne, "The Cooling World," *Newsweek*, April 28, 1975,
http://www.scribd.com/doc/225798861/Newsweek-s-Global
-Cooling-Article-From-April-28-1975.

17. P. Gwynne, "My 1975 'Cooling World' Story Doesn't Make
Today's Climate Scientists Wrong," Inside Science, May 21, 2014,

https://www.insidescience.org/content/my-1975-cooling-world
-story-doesnt-make-todays-climate-scientists-wrong/1640.

18. "Another Ice Age?" *Time*, June 24, 1974, http://content.time.com/
time/magazine/article/0,9171,944914,00.html.

19. T. C. Peterson, W. M. Connolley, and J. Fleck, "The Myth of the
1970s Global Cooling Scientific Consensus," *Bulletin of the American Meteorological Society* 89 (2008): 1325–37, http://journals.ametsoc
.org/doi/abs/10.1175/2008BAMS2370.1.

20. National Research Council, Climate Research Board, "Carbon
Dioxide and Climate: A Scientific Assessment," July 1979, http://
web.atmos.ucla.edu/~brianpm/download/charney_report.pdf.

21. Rick Perry, "Statement by Gov. Perry on Planned Parenthood
Video," Perry for President, July 14, 2015, https://rickperry.org/
planned-parenthood-video.

22. Carly Fiorina, Facebook post, July 14, 2015, https://www.face
book.com/CarlyFiorina/posts/10156174378690206.

23. Rand Paul, Twitter tweet, July 14, 2015, https://twitter.com/Rand
Paul/status/621118379709980673.

24. American Society for Cell Biology, "Talking Points: Fetal Tissue Research," April 2001, https://web.archive.org/web/20070713
032718/http://www.ascb.org/newsfiles/fetaltissue.pdf.

25. "The Nobel Prize in Physiology or Medicine 1954," Nobelprize
.org, http://www.nobelprize.org/nobel_prizes/medicine/laureates/
1954, accessed April 13, 2016.

26. Smithsonian National Museum of American History, "History
of Vaccines," http://amhistory.si.edu/polio/virusvaccine/history3
.htm, accessed April 13, 2016.

27. National Institutes of Health Revitalization Act of 1993, 103rd
Congr. (1993), http://www.gpo.gov/fdsys/pkg/BILLS-103s1enr/
pdf/BILLS-103s1enr.pdf.

28. D. Levitan, "Unspinning the Planned Parenthood Video," Fact-
Check.org, July 21, 2015, http://www.factcheck.org/2015/07/
unspinning-the-planned-parenthood-video. Quotes provided to
the author while employed at FactCheck.org.

29. "Human Fetal Tissue: Acquisition for Federally Funded Biomedical Research," US General Accounting Office, October 4, 2000,
http://www.gao.gov/products/164170.

30. N. Pemberton, "Missouri Joins Growing List of States Unable to

Pin Anything on Planned Parenthood," *New York Magazine*, September 29, 2015, http://nymag.com/daily/intelligencer/2015/09/state-has-found-proof-of-fetal-tissue-sales.html.

31. R. T. Beckwith, "Transcript: Read the Full Text of the Second Republican Debate," Time, September 16, 2015, http://time.com/4037239/second-republican-debate-transcript-cnn.

32. StemExpress, "Statement of StemExpress concerning Recent Media Stories," July 16, 2015, http://stemexpress.com/statement-of-stemexpress-concerning-recent-media-stories.

33. M. Scherer, "Questions Remain over Whether Video Carly Fiorina Cited in Debate Shows Abortion," Time, September 29, 2015, http://time.com/4055143/abortion-carly-fiorina-planned-parenthood-2.

34. Associated Press, "Suspect in Planned Parenthood Attack Said 'No More Baby Parts' after Arrest," *Guardian*, November 28, 2015, http://www.theguardian.com/us-news/2015/nov/29/suspect-in-planned-parenthood-attack-said-no-more-baby-parts-after-arrest.

35. P. Vercammen and H. Yan, "Robert Dear Has Outbursts at Hearing," CNN, December 9, 2015, http://www.cnn.com/2015/12/09/us/colorado-planned-parenthood-shooting.

CHAPTER 6: THE RIDICULE AND DISMISS

1. "Iowa Freedom Summit, Mike Huckabee," C-SPAN, January 24, 2015, http://www.c-span.org/video/?323834-17/iowa-freedom-summit-mike-huckabee.

2. White House, Office of the Press Secretary, "Remarks by the President in State of the Union Address," January 20, 2015, https://www.whitehouse.gov/the-press-office/2015/01/20/remarks-president-state-union-address-january-20-2015.

3. Information on these gases and their warming potential is available here: Environmental Protection Agency, "Overview of Greenhouse Gases," http://www.epa.gov/climatechange/ghgemissions/gases/fgases.html, accessed April 20, 2016.

4. US Department of Defense, "2014 Climate Change Adaptation Roadmap," https://www.scribd.com/doc/242845848/Read-DoD-report-2014-Climate-Change-Adaptation-Roadmap, accessed April 20, 2016.

5. C. P. Kelley, S. Mohtadi, M. A. Cane, et al., "Climate Change

in the Fertile Crescent and Implications of the Recent Syrian Drought," *Proceedings of the National Academy of Sciences* 112 (2015): 3241–46, http://www.pnas.org/content/112/11/3241.full.

6. T. Coburn, "WasteBook 2012," October 2012, http://showmethe spending.com/wp-content/uploads/2014/06/Wastebook2012.pdf.

7. "Senator Rand Paul Remarks at *American Spectator* Gala," February 11, 2015, C-Span, http://www.c-span.org/video/?324313-1/senator-rand-paul-rky-remarks-american-spectator-gala.

8. Pletcher Lab, https://sites.google.com/a/umich.edu/pletcher-lab.

9. C. M. Gendron, T. H. Kuo, Z. M. Harvanek, et al., "*Drosophila* Life Span and Physiology Are Modulated by Sexual Perception and Reward," *Science* 343 (2013): 544–48, http://www.ncbi.nlm.nih.gov/pmc/articles/PMC4042187.

10. University of Michigan Health System, "Fruit Flies with Better Sex Lives Live Longer," EurekAlert!, November 28, 2013, http://www.eurekalert.org/pub_releases/2013-11/uomh-ffw112513.php.

11. T.-H. Kuo, J. Y. Yew, T. Y. Fedina, et al., "Aging Modulates Cuticular Hydrocarbons and Sexual Attractiveness in *Drosophila melanogaster*," *Journal of Experimental Biology* 215 (2012): 814–21, http://jeb.biologists.org/content/215/5/814.

12. N. Gingrich, "Double the NIH Budget," *New York Times*, April 22, 2015, http://www.nytimes.com/2015/04/22/opinion/double-the-nih-budget.html?_r=0.

13. D. Levitan, "Paul Knocks Flies and NIH Funding," FactCheck.org, February 19, 2015, http://www.factcheck.org/2015/02/paul-knocks-flies-and-nih-funding. Quotes provided to the author while employed at FactCheck.org.

14. Estimates of total genes in various organisms are available at Leslie A. Pray, "Eukaryotic Genome Complexity," Scitable by Nature Education, 2008, http://www.nature.com/scitable/topicpage/eukaryotic-genome-complexity-437.

15. I. Miko, "Thomas Hunt Morgan and Sex Linkage," Scitable by Nature Education, 2008, http://www.nature.com/scitable/topicpage/Thomas-Hunt-Morgan-and-Sex-Linkage-452.

16. "The Nobel Prize in Physiology or Medicine 1933," Nobelprize.org, http://www.nobelprize.org/nobel_prizes/medicine/laureates/1933, accessed April 20, 2016.

17. "The Nobel Prize in Physiology or Medicine 1946," Nobelprize.org,

http://www.nobelprize.org/nobel_prizes/medicine/laureates/1946, accessed April 20, 2016.

18. "The Nobel Prize in Physiology or Medicine 1995," Nobelpriz.org, http://www.nobelprize.org/nobel_prizes/medicine/laureates/1995, accessed April 20, 2016; "The Nobel Prize in Physiology or Medicine 2011," Nobelprize.org, http://www.nobelprize.org/nobel_prizes/medicine/laureates/2011, accessed April 20, 2016.

19. Levitan, "Paul Knocks Flies and NIH Funding."

20. R. Irion, "What Proxmire's Golden Fleece Did for—and to—Science," *Scientist*, December 12, 1988, http://www.the-scientist.com/?articles.view/articleNo/10030/title/What-Proxmire-s-Golden-fleece-Did-For--And-To--Science.

21. "2014: Rat and Infant Massage," Golden Goose Award, 2014, http://www.goldengooseaward.org/awardees/2014/12/4/2014-rat-and-infant-massage, accessed April 20, 2016.

CHAPTER 7: THE LITERAL NITPICK

1. US Senate Committee on Environment & Public Works, "Inhofe Calls Hydraulic Fracturing Rule 'Unnecessary,' Rejects Rule with Introduction of FRESH Act," March 20, 2015, http://www.epw.senate.gov/public/index.cfm/press-releases-republican?ID=31C63566-C7A1-79FE-3F39-761A806C1AD2.

2. Bureau of Land Management, "Interior Department Releases Final Rule to Support Safe, Responsible Hydraulic Fracturing Activities on Public and Tribal Lands," March 20, 2015, http://www.blm.gov/wo/st/en/info/newsroom/2015/march/nr_03_20_2015.html.

3. FracFocus Chemical Disclosure Registry, "What Chemicals Are Used," https://fracfocus.org/chemical-use/what-chemicals-are-used, accessed April 20, 2016.

4. Centers for Disease Control and Prevention, "Facts about Benzene," http://www.bt.cdc.gov/agent/benzene/basics/facts.asp, accessed April 20, 2016.

5. Pennsylvania Department of Environmental Protection, "Water Supply Determination Letters," http://files.dep.state.pa.us/OilGas/BOGM/BOGMPortalFiles/OilGasReports/Determination_Letters/Regional_Determination_Letters.pdf, accessed April 20, 2016.

6. T. H. Darrah, A. Vengosh, R. B. Jackson, et al., "Noble Gases Identify the Mechanisms of Fugitive Gas Contamination in Drinking-Water Wells Overlying the Marcellus and Barnett Shales," *Proceedings of the National Academy of Sciences* 111 (2014): 14076–81, http://www.pnas.org/content/111/39/14076.long.

7. US Geological Survey, "Hydraulic Fracturing Fluids Likely Harmed Threatened Kentucky Fish Species," August 28, 2013, http://www.usgs.gov/newsroom/article.asp?ID=3677#.VlMmimSrRN2.

8. D. M. Papoulias and A. L. Velasco, "Histopathological Analysis of Fish from Acorn Fork Creek, Kentucky, Exposed to Hydraulic Fracturing Fluid Releases," *Southeastern Naturalist* 12 (2013): 92–111, http://www.eaglehill.us/SENAonline/articles/SENA-sp-4/18-Papoulias.shtml.

9. B. E. Fontenot, L. R. Hunt, Z. L. Hildenbrand, et al., "An Evaluation of Water Quality in Private Drinking Water Wells near Natural Gas Extraction Sites in the Barnett Shale Formation," *Environmental Science & Technology* 47 (2013): 10032–40, http://pubs.acs.org/doi/abs/10.1021/es4011724.

10. Environmental Protection Agency, "Assessment of the Potential Impacts of Hydraulic Fracturing for Oil and Gas on Drinking Water Resources: Executive Summary," June 2015, http://www.epa.gov/sites/production/files/2015-06/documents/hf_es_erd_jun2015.pdf.

11. US Fish and Wildlife Service, Environmental Conservation Online System, "Blackside Dace (*Phoxinus cumberlandensis*)," http://ecos.fws.gov/tess_public/profile/speciesProfile.action?spcode=E05I, accessed April 20, 2016.

12. Donelle Harder (communications director for Senator James Inhofe), e-mail to the author, March 24, 2015.

13. Paula Reid and Stephanie Condon, "DEA Chief Says Smoking Marijuana as Medicine 'Is a Joke,'" CBS News, November 4, 2015, http://www.cbsnews.com/news/dea-chief-says-smoking-marijuana-as-medicine-is-a-joke.

14. Food and Drug Administration, "FDA and Marijuana: Questions and Answers," http://www.fda.gov/NewsEvents/PublicHealthFocus/ucm421168.htm#notapproved, accessed April 20, 2016.

15. Drug Enforcement Administration, "Drug Scheduling," http://www.dea.gov/druginfo/ds.shtml, accessed April 20, 2016.

16. D. I. Abrams, C. A. Jay, S. B. Shade, et al., "Cannabis in Painful HIV-Associated Sensory Neuropathy: A Randomized Placebo -Controlled Trial," *Neurology* 68 (2007): 515–21, http://www.ncbi .nlm.nih.gov/pubmed/17296917.

17. R. J. Ellis, W. Toperoff, F. Vaida, et al., "Smoked Medicinal Cannabis for Neuropathic Pain in HIV: A Randomized, Crossover Clinical Trial," *Neuropsychopharmacology* 34 (2009): 672–80, http:// www.nature.com/npp/journal/v34/n3/full/npp2008120a.html.

18. J. Corey-Bloom, T. Wolfson, A. Gamst, et al., "Smoked Cannabis for Spasticity in Multiple Sclerosis: A Randomized, Placebo -Controlled Trial," *Canadian Medical Association Journal* 184 (2012): 1143–50, http://www.cmaj.ca/content/184/10/1143.long.

CHAPTER 8: THE CREDIT SNATCH

1. "Rick Perry Remarks at CPAC," C-SPAN, February 27, 2015, http://www.c-span.org/video/?324558-4/former-governor-rick -perry-rtx-remarks-cpac-2015.

2. Texas Commission on Environmental Quality, "Texas Statewide Point Source NO_x Emissions Trends," http://www.tceq.state .tx.us/assets/public/implementation/air/success/successImages/ inventory/txNOxTrend.png, accessed April 20, 2016.

3. Texas Commission on Environmental Quality, "2014 NO_x Emissions in Texas," https://tceq.state.tx.us/airquality/areasource/emis sions-sources-charts/2011texasNOXemissions/view, accessed April 20, 2016.

4. Environmental Protection Agency, "1970–2014 Average Annual Emissions, All Criteria Pollutants," National Emissions Inventory, March 2015, https://www.epa.gov/sites/production/files/2015-07/ national_tier1_caps.xlsx.

5. Energy Information Administration, "Power Plant Emissions of Sulfur Dioxide and Nitrogen Oxides Continue to Decline in 2012," February 27, 2013, http://www.eia.gov/todayinenergy/detail .cfm?id=10151.

6. After Perry's comments, the EIA updated its data, in a way that showed a sharply lower reduction in CO_2. Perry's numbers were correct at the time he mentioned them, and this analysis focuses on those earlier figures.

7. Environmental Protection Agency, "Sources of Greenhouse Gas Emissions," http://www3.epa.gov/climatechange/ghgemissions/sources/industry.html, accessed April 20, 2016.

8. "Greenhouse Gas Inventory," *Climate Action Report*, 2014, 82, http://www.state.gov/documents/organization/218991.pdf.

9. American Wind Energy Association, "U.S. Wind Energy State Facts," data current to end of fourth quarter 2015, http://www.awea.org/resources/statefactsheets.aspx?itemnumber=890, accessed April 20, 2016.

10. WINDExchange, "Installed Wind Capacity," US Department of Energy, http://apps2.eere.energy.gov/wind/windexchange/wind_installed_capacity.asp#yearly, accessed April 20, 2016.

11. R. R. Drouin, "How Conservative Texas Took the Lead in US Wind Power," *Yale Environment 360*, April 9, 2015, http://e360.yale.edu/feature/how_conservative_texas_took_the_lead_in_us_wind_power/2863.

12. "Renewable Generation Requirement: Program Overview" (Texas), Database of State Incentives for Renewables & Efficiency, http://programs.dsireusa.org/system/program/detail/182, accessed April 20, 2016.

13. D. Matthews, "Rick Perry's Environmental Record," *Washington Post*, August 15, 2011. https://www.washingtonpost.com/blogs/ezra-klein/post/rick-perrys-environmental-record/2011/08/15/gIQApJbzGJ_blog.html.

14. "Chris Christie at the Iowa Ag Summit," YouTube, March 16, 2015, https://www.youtube.com/watch?v=6xEIXMPFEus.

15. Solar Energy Industries Association, "New Jersey Solar," http://www.seia.org/state-solar-policy/New-Jersey, accessed April 20, 2016.

16. "Renewables Portfolio Standard: Program Overview" (New Jersey), Database of State Incentives for Renewables & Efficiency, http://programs.dsireusa.org/system/program/detail/564, accessed April 20, 2016.

17. Energy Information Administration, "FAQ: How Much Electricity Does an American Home Use?" https://www.eia.gov/tools/faqs/faq.cfm?id=97&t=3, accessed April 20, 2016.

18. M. Rao, "Christie Rolls Out Energy Plan, Rolls Back Renewable Energy Goals," *Philadelphia Inquirer*, June 8, 2011, http://

articles.philly.com/2011-06-08/news/29634207_1_renewable
-energy-energy-plan-energy-future.

19. E. Wesoff, "Gov. Christie Signs New Jersey's Solar 'Resurrection Bill,'" Greentech Media, July 23, 2012, http://www.greentechmedia .com/articles/read/New-Jerseys-Solar-Resurrection-Bill-Passes.

20. S. P. Sullivan, "Christie Administration Joins Effort to Block Obama's Clean Power Plan," NJ.com, October 23, 2015, http:// www.nj.com/politics/index.ssf/2015/10/christie_administration_ joins_effort_to_block_obam.html.

21. Regional Greenhouse Gas Initiative, "Welcome," http://www .rggi.org, accessed April 20, 2016.

22. C. Maza, "Everyone's Favorite Climate Change Fix," *Christian Science Monitor*, October 29, 2015, http://www.csmonitor.com/ Environment/Energy/2015/1029/Everyone-s-favorite-climate -change-fix.

23. White House, Office of the Press Secretary, "President Bush Commemorates Earth Day 2007, Calls for Good Stewardship of Land and Oceans," April 20, 2007, https://web.archive .org/web/20070510012518/http://www.whitehouse.gov/news/ releases/2007/04/20070420-8.html.

24. White House, "Energy for America's Future," http://georgewbush- whitehouse.archives.gov/infocus/energy, accessed April 20, 2016.

25. L. Tillemann, F. Beck, J. Brodrick, et al., "Revolution Now: The Future Arrives for Four Clean Energy Technologies," US Department of Energy, September 17, 2013, http://energy.gov/sites/prod/ files/2013/09/f2/200130917-revolution-now.pdf.

26. Office of Energy Efficiency & Renewable Energy, "About the SunShot Initiative," http://energy.gov/eere/sunshot/about-sunshot -initiative, accessed April 20, 2016.

27. Union of Concerned Scientists, "Bush Budget Slashes Funds for Renewable Energy Sources," February 6, 2003, https://web .archive.org/web/20060602144949/http://www.ucsusa.org/news/ press_release/bush-budget-slashes-funds-for-renewable-energy -sources.html.

28. Union of Concerned Scientists, "Bush Administration FY06 Budget—Highlights and Lowlights," February 22, 2005, https:// web.archive.org/web/20060615033724/http://www.ucsusa.org/ news/positions/president-bushs-fy-2006-budget.html.

29. "Energy Shortage" (Editorial), *New York Times*, July 28, 2005, http://www.nytimes.com/2005/07/28/opinion/energy-shortage.html.

30. J. Blum, "House Energy Bill Increases Tax Breaks," *Washington Post*, April 19, 2005, http://www.washingtonpost.com/wp-dyn/articles/A63958-2005Apr18.html.

31. D. J. Weiss, "Renewable Energy Subterfuge: Bush's Sleight of Hand," Center for American Progress, March 4, 2008, https://www.americanprogress.org/issues/green/news/2008/03/04/4048/renewable-energy-subterfuge-bushs-sleight-of-hand.

CHAPTER 9: THE CERTAIN UNCERTAINTY

1. Commission on Presidential Debates, "October 11, 2000 Debate Transcript," http://www.debates.org/?page=october-11-2000-debate-transcript.

2. S. Lewandowsky, "Long Term Certainty," Skeptical Science, August 18, 2010, http://www.skepticalscience.com/Long-Term-Certainty.html.

3. C. Dann, "Jeb Bush Blasts 'Intellectual Arrogance' in Climate Change Debate," NBC News, May 21, 2015, http://www.nbcnews.com/politics/2016-election/jeb-bush-blasts-intellectual-arrogance-climate-change-debate-n362586.

4. N. L. Bindoff, K. M. AchutaRao, M. R. Allen, et al., "Detection and Attribution of Climate Change: From Global to Regional," in *Climate Change 2013: The Physical Science Basis. Contribution of Working Group I to the Fifth Assessment Report of the Intergovernmental Panel on Climate Change*, ed. T. F. Stocker, D. Qin, G.-K. Plattner, et al. (Cambridge University Press, 2013), 887, https://www.ipcc.ch/pdf/assessment-report/ar5/wg1/WG1AR5_Chapter10_FINAL.pdf.

5. R. Lindsey, "If Earth Has Warmed and Cooled throughout History, What Makes Scientists Think That Humans Are Causing Global Warming Now?" *Earth Observatory* (blog), NASA, May 4, 2010, http://earthobservatory.nasa.gov/blogs/climateqa/if-earth-has-warmed-and-cooled-throughout-history-what-makes-scientists-think-that-humans-are-causing-global-warming-now.

6. Bindoff et al., "Detection and Attribution of Climate Change," 884 (fig. 10.5).

7. T. M. L. Wigley and B. D. Santer, "A Probabilistic Quantification of the Anthropogenic Component of Twentieth Century Global Warming," *Climate Dynamics* 40 (2012): 1087–1102, http://link.springer.com/article/10.1007%2Fs00382-012-1585-8.

8. M. Huber and R. Knutt, "Anthropogenic and Natural Warming Inferred from Changes in Earth's Energy Balance," *Nature Geoscience* 5 (2012): 31–36, http://www.nature.com/ngeo/journal/v5/n1/abs/ngeo1327.html.

9. Marco Rubio, on *Face the Nation*: "Face the Nation Transcripts, April 19, 2015: Rubio, Manchin, O'Malley," CBS News, http://www.cbsnews.com/news/face-the-nation-transcripts-april-19-2015-rubio-manchin-omalley.

10. N. Stern, "Stern Review: The Economics of Climate Change: Executive Summary," http://siteresources.worldbank.org/INTINDONESIA/Resources/226271-1170911056314/3428109-1174614780539/SternReviewEng.pdf, accessed April 20, 2016.

11. M. Z. Jacobson, M. A. Delucchi, M. A. Cameron, et al., "Low-Cost Solution to the Grid Reliability Problem with 100% Penetration of Intermittent Wind, Water, and Solar for All Purposes," *Proceedings of the National Academy of Sciences* 112 (2015): 15060–65, http://www.pnas.org/content/early/2015/11/18/1510028112.

12. B. Carey, "Stanford Engineers Develop State-by-State Plan to Convert U.S. to 100% Clean, Renewable Energy by 2050," Stanford News Service, June 8, 2015, http://cee.stanford.edu/news-events/news/stanford-engineers-develop-state-state-plan-convert-us-100-clean-renewable-energy.

13. M. Burke, S. M. Hsiang, and E. Miguel, "Global Non-linear Effect of Temperature on Economic Production," *Nature* 527 (2015): 235–39, http://www.nature.com/nature/journal/v527/n7577/full/nature15725.html.

14. J. Worland, "Climate Change Could Wreck the Global Economy," *Time*, October 22, 2015, http://time.com/4082328/climate-change-economic-impact.

15. National Economic Research Associates, "Potential Energy Impacts of the EPA Proposed Clean Power Plan," October 17, 2014, http://

www.nera.com/content/dam/nera/publications/2014/NERA_
ACCCE_CPP_Final_10.17.2014.pdf.

16. J. Rogers, "ACCCE, NERA, and Another Misleading Study
about the Clean Power Plan," *The Equation* (blog), Union of
Concerned Scientists, November 12, 2015, http://blog.ucsusa.org/
john-rogers/accce-nera-and-another-misleading-study-about-the
-clean-power-plan-952.

17. Environmental Protection Agency, "Regulatory Impact Analysis for the Clean Power Plan Final Rule," http://www.epa.gov/
sites/production/files/2015-08/documents/cpp-final-rule-ria.pdf,
accessed April 20, 2016.

18. Citi, "Energy Darwinism II: Why a Low Carbon Future Doesn't
Have to Cost the Earth," August 2015, https://ir.citi.com/hsq32J
l1m4aIzicMqH8sBkPnbsqfnwy4JgbIJ2kIPYWIw5eM8yD3FY9V
bGpK%2Baax.

19. Environmental Protection Agency, "Biography of Lee M. Thomas,"
http://www.epa.gov/aboutepa/biography-lee-m-thomas,
accessed April 20, 2016.

20. M. Borzilleri, "More Study Needed on Acid Rain Problem,"
Adirondack Daily Enterprise, August 21, 1985, http://nyshistoricnews
papers.org/lccn/sn86033360/1985-08-21/ed-1/seq-1.pdf.

21. Associated Press, "Reagan Position on Acid Rain Study Is 'Incredible,' Coloradan Charges," April 30, 1986, https://news.google.com/
newspapers?nid=336&dat=19860430&id=NkxTAAAAIBAJ&sjid
=3YMDAAAAIBAJ&pg=3794,6433344&hl=en.

22. B. A. Franklin, "Legislators Sat [*sic*] White House Suppressed
Acid Rain Report," *New York Times*, August 17, 1984, http://
www.nytimes.com/1984/08/18/us/legislators-sat-white-house
-suppressed-acid-rain-report.html.

23. National Research Council, *Acid Deposition: Atmospheric Processes
in Eastern North America: A Review of Current Scientific Understanding*
(National Academy Press, 1983), http://www.nap.edu/read/182/
chapter/1.

24. R. Blumenthal, "Texas Legislators Block Shots for Girls against
Cancer Virus," *New York Times*, April 26, 2007, http://www
.nytimes.com/2007/04/26/us/26texas.html.

25. Centers for Disease Control and Prevention, "Genital HPV

Infection—Fact Sheet," http://www.cdc.gov/std/hpv/stdfact-hpv
.htm, accessed April 20, 2016.

26. D. Grady, "Vaccine Prevents Most Cervical Cancer," *New York Times*, October 7, 2005, http://www.nytimes.com/2005/10/07/
health/vaccine-prevents-most-cervical-cancer.html.

27. The full results of this trial were published in the *New England Journal of Medicine* in May 2007 [Future II Study Group, "Quadri-valent Vaccine against Human Papillomavirus to Prevent High-Grade Cervical Lesions," *New England Journal of Medicine* 356 (2007): 1915–27, http://www.nejm.org/doi/full/10.1056/NEJMoa061741
#t=articleResults], a few days after the Texas controversy. The study made news earlier, though, when presented at an interna-tional infectious diseases conference. Medical findings are often first presented at conferences and then published in peer-reviewed journals.

28. P. Moyer, "Two HPV Vaccines Show Feasibility of Vaccinat-ing Young Girls," Medscape Medical News, December 19, 2005, http://www.medscape.com/viewarticle/520121.

29. This study was presented at a conference in December 2005 and published in the *Journal of Adolescent Health* [C. Pedersen, T. Petaja, G. Strauss, et al., "Immunization of Early Adolescent Females with Human Papillomavirus Type 16 and 18 L1 Virus-like Parti-cle Vaccine Containing AS04 Adjuvant," 40 (2007): 564–71, http://
www.jahonline.org/article/S1054-139X(07)00106-1/abstract] soon after the Texas legislature vote, in June 2007. Again, the evidence was available at the time of the vote.

30. G. Harris, "Panel Unanimously Recommends Cervical Can-cer Vaccine for Girls 11 and Up," *New York Times*, June 30, 2006, http://www.nytimes.com/2006/06/30/health/30vaccine.html.

31. Centers for Disease Control and Prevention, "HPV Vaccines: Vaccinating Your Preteen or Teen," http://www.cdc.gov/hpv/
parents/vaccine.html, accessed April 20, 2016.

CHAPTER 10: THE BLIND EYE TO FOLLOW-UP

1. H. Olivecrona, "The Nobel Prize in Physiology or Medicine 1949: Award Ceremony Speech," http://www.nobelprize.org/

nobel_prizes/medicine/laureates/1949/press.html, accessed April 25, 2016.

2. White House, Office of the Press Secretary, "Remarks by the President on Precision Medicine," January 30, 2015, https://www.whitehouse.gov/the-press-office/2015/01/30/remarks-president-precision-medicine.

3. Battelle Memorial Institute, "About Us," http://www.battelle.org/about-us, accessed April 25, 2016.

4. S. Tripp and M. Grueber, "Economic Impact of the Human Genome Project," May 2011, http://www.battelle.org/docs/default-document-library/economic_impact_of_the_human_genome_project.pdf.

5. N. Drake, "What Is the Human Genome Worth?" *Nature*, May 11, 2011, http://www.nature.com/news/2011/110511/full/news.2011.281.html.

6. Battelle Technology Partnership Practice, "The Impact of Genomics on the U.S. Economy," June 2013, http://web.ornl.gov/sci/techresources/Human_Genome/publicat/2013BattelleReportImpact-of-Genomics-on-the-US-Economy.pdf.

7. White House, Office of the Press Secretary, "Remarks by the President on the BRAIN Initiative and American Innovation," April 2, 2013, https://www.whitehouse.gov/the-press-office/2013/04/02/remarks-president-brain-initiative-and-american-innovation.

8. Drake, "What Is the Human Genome Worth?"

9. Gary Palmer, interviewed on the Matt Murphy Show (no longer accessible online). Originally transcribed by the author and printed in the following story: D. Levitan, "Nothing False about Temperature Data," FactCheck.org, February 12, 2015, http://www.factcheck.org/2015/02/nothing-false-about-temperature-data.

10. T. J. Zinser (US Department of Commerce Inspector General) to Senator James M. Inhofe, February 18, 2011, https://www.oig.doc.gov/OIGPublications/2011.02.18-IG-to-Inhofe.pdf.

11. R. Oxburgh, H. Davies, K. Emanuel, et al., "Report of the International Panel Set Up by the University of East Anglia to Examine the Research of the Climatic Research Unit," April 12, 2010, http://www.uea.ac.uk/documents/3154295/7847337/SAP.pdf/a6f591fc-fc6e-4a70-9648-8b943d84782b.

12. Environmental Protection Agency, "Myths vs. Facts: Denial of Petitions for Reconsideration of the Endangerment and Cause or Contribute Findings for Greenhouse Gases under Section 202(a) of the Clean Air Act," http://www3.epa.gov/climatechange/endan germent/myths-facts.html, accessed April 25, 2016.

13. T. Gettys, "Global Warming Denier Jim Inhofe: 'Fewer and Fewer' Senators Believe in Climate Change 'Hoax,'" Raw Story, January 13 2014, http://www.rawstory.com/2014/01/global-warming -denier-jim-inhofe-fewer-and-fewer-senators-believe-in-climate -change-hoax.

14. J. M. Inhofe, "Inhofe: 'Climategate' Should Have Ended the Global Warming Debate" (speech, Heritage Foundation, June 2015), YouTube, June 24, 2015, https://www.youtube.com/watch ?v=3q3B5enm9DY.

15. W. Saletan, "Unhealthy Fixation," Slate, July 15, 2015, http:// www.slate.com/articles/health_and_science/science/2015/07/are _gmos_safe_yes_the_case_against_them_is_full_of_fraud_lies _and_errors.html.

16. Food and Drug Administration, "FDA Takes Several Actions Involving Genetically Engineered Plants and Animals for Food," November 19, 2015, http://www.fda.gov/NewsEvents/Newsroom/ PressAnnouncements/ucm473249.htm.

17. Senator Lisa Murkowski, "Alaska Delegation Responds to FDA Approval of GE Salmon," November 19, 2015, http://www .murkowski.senate.gov/public/index.cfm/pressreleases?ID=a285c 8c9-1d48-44f0-ab98-9688d68821e5.

18. Ibid.

19. Jared Huffman, "Huffman Blasts FDA Approval of Genetically -Engineered Salmon," November 19, 2015, https://huffman.house .gov/media-center/press-releases/huffman-blasts-fda-approval-of -genetically-engineered-salmon.

20. A. Pollack, "Genetically Engineered Salmon Approved for Consumption," New York Times, November 19, 2015, http://www.ny times.com/2015/11/20/business/genetically-engineered-salmon -approved-for-consumption.html?_r=0.

21. Food and Drug Administration, "AquAdvantage Salmon Fact Sheet," http://www.fda.gov/AnimalVeterinary/DevelopmentAp

provalProcess/GeneticEngineering/GeneticallyEngineered Animals/ucm473238.htm, accessed April 25, 2016.

22. Alaska Department of Fish and Game, "Commercial Fisheries," http://www.adfg.alaska.gov/index.cfm?adfg=fishingcommercial .main, accessed April 25, 2016.

23. World Health Organization, "Frequently Asked Questions on Genetically Modified Foods," http://www.who.int/foodsafety/ areas_work/food-technology/faq-genetically-modified-food/en, accessed April 25, 2016.

CHAPTER 11: THE LOST IN TRANSLATION

1. Environmental Protection Agency, "Estimated Per Capita Fish Consumption in the United States," August 2002, http://water .epa.gov/scitech/swguidance/standards/criteria/health/upload/ consumption_report.pdf.

2. InsideSources, "Road to 2016: Des Moines, Iowa," YouTube, April 10, 2015, https://www.youtube.com/watch?v=vi00XFzKb2g.

3. Environmental Protection Agency, "Regulatory Impact Analysis for the Final Mercury and Air Toxics Standards," December 2011, http://www3.epa.gov/ttn/ecas/regdata/RIAs/matsriafinal.pdf.

4. The full methodology used to determine the exposed population is available in Environmental Protection Agency, "Regulatory Impact Analysis for the Final Mercury and Air Toxics Standards."

5. US Department of the Interior, Fish and Wildlife Service, and US Department of Commerce, Census Bureau, "2006 National Survey of Fishing, Hunting, and Wildlife-Associated Recreation," https://www.census.gov/prod/2008pubs/fhw06-nat.pdf.

6. B. Potts, "Mercurial Regulators Making Fishy Calculations," *Wall Street Journal*, March 23, 2015, http://www.wsj.com/articles/ brian-potts-mercurial-regulators-making-fishy-calculations-14 27153569.

7. B. Potts, e-mails to and phone conversations with the author, April 14–17, 2015, while employed at FactCheck.org.

8. "State of Michigan et al., Petitioners, v. Environmental Protection Agency, Respondent: Brief for the Cato Institute as *Amicus Curiae* in Support of Petitioners," January 2015, http://www

.americanbar.org/content/dam/aba/publications/supreme_court_
preview/BriefsV5/14-46_amicus_pet_cato.authcheckdam.pdf.

9. Environmental Protection Agency, "Regulatory Impact Analysis
for the Final Mercury and Air Toxics Standards," 4-39.

10. "Dan Bongino and Rand Paul, Baltimore County Lincoln Day
Dinner, Full Speeches," YouTube, June 11, 2015, https://www.you
tube.com/watch?v=1l-_t_4Z5h8&feature=youtu.be&t=16m26s.

11. Environmental Protection Agency, "CWA Section 404 Enforcement Overview," http://www.epa.gov/cwa-404/cwa-section-404
-enforcement-overview, accessed April 25, 2016.

12. Environmental Protection Agency, "Section 404 of the Clean
Water Act: How Wetlands are Defined and Identified," http://
www.epa.gov/cwa-404/section-404-clean-water-act-how-wet
lands-are-defined-and-identified, accessed April 25, 2016.

13. US Department of Justice, "Grand Jury in Jackson, Mississippi,
Charges Three Individuals and Two Corporations with Wetlands
Violations," June 10, 2004, http://www.justice.gov/archive/opa/
pr/2004/June/04_enrd_396.htm.

14. W. Emmerich, "More on Wetlands Injustice," *Enterprise-Journal*,
August 1, 2008, http://www.enterprise-journal.com/article_b62a6
6a2-fb84-5706-9b63-ee3e16999db5.html; W. Emmerich, "A Nightmare Is Over," *Enterprise-Journal*, September 30, 2010, http://www
.enterprise-journal.com/opinion/article_2db7a8e6-ccc9-11df-a462
-001cc4c03286.html.

15. US Department of the Interior, "Five-Year Survey Shows Wetlands Losses Are Slowing, Marking Conservation Gains and
Need for Continued Investment in Habitat," October 6, 2011,
http://www.fws.gov/wetlands/Documents/Status-and-Trends
-of-Wetlands-in-the-Conterminous-United-States-2004-to-2009
-News-Release.pdf.

16. Ibid.

CHAPTER 12: THE STRAIGHT-UP FABRICATION

1. C. Jaco, "Jaco Report: Full Interview with Todd Akin," Fox 2,
St. Louis, August 19, 2012, http://fox2now.com/2012/08/19/the
-jaco-report-august-19-2012.

2. M. P. Jeffries, "How Rap Can Help End Rape Culture," *Atlan-*

tic, October 30, 2012, http://www.theatlantic.com/entertainment/archive/2012/10/how-rap-can-help-end-rape-culture/264258.

3. E. Bradner and A. Jaffe, "Ben Carson Apologizes for Comments on Gay People," CNN, March 5, 2015, http://www.cnn.com/2015/03/04/politics/ben-carson-prisons-gay-choice.

4. American Psychological Association, "Sexual Orientation & Homosexuality: Answers to Your Questions for a Better Understanding," http://www.apa.org/topics/lgbt/orientation.aspx, accessed April 25, 2016.

5. G. M. Herek, A. T. Norton, T. J. Allen, et al., "Demographic, Psychological, and Social Characteristics of Self-Identified Lesbian, Gay, and Bisexual Adults in a US Probability Sample," *Sexuality Research & Social Policy* 7 (2010): 176–200, http://www.ncbi.nlm.nih.gov/pmc/articles/PMC2927737.

6. D. H. Hamer, S. Hu, V. L. Magnuson, et al., "A Linkage between DNA Markers on the X Chromosome and Male Sexual Orientation," *Science* 261 (1993): 321–27, http://www.sciencemag.org/content/261/5119/321.long.

7. W. R. Rice, U. Fribert, and S. Gavrilets, "Homosexuality as a Consequence of Epigenetically Canalized Sexual Development," *Quarterly Review of Biology* 87 (2012): 343–68, http://www.jstor.org/stable/10.1086/668167.

8. A. Camperio-Ciani, F. Corna, and C. Capiluppi, "Evidence for Maternally Inherited Factors Favouring Male Homosexuality and Promoting Female Fecundity," *Proceedings of the Royal Society B: Biological Sciences*, 271 (2004): 2217–21, http://rspb.royalsocietypublishing.org/content/271/1554/2217.long.

9. A. Camperio Ciani and E. Pellizzari, "Fecundity of Paternal and Maternal Non-parental Female Relatives of Homosexual and Heterosexual Men," *PLoS One* 7 (2012): e51088, http://journals.plos.org/plosone/article?id=10.1371/journal.pone.0051088.

10. D. Levitan, "Carson's Missteps on Sexual Orientation," FactCheck.org, March 6, 2015, http://www.factcheck.org/2015/03/carsons-missteps-on-sexual-orientation. Quotes provided to the author in a phone interview while employed at FactCheck.org.

11. M. Walsh, "Mike Huckabee Stands behind Holocaust Comment as Backlash Continues," July 28, 2015, https://www.yahoo.com/politics/mike-huckabee-stands-behind-holocaust-comment-as

-125266374196.html, including Yahoo! News Live interview with Huckabee.

12. US Geological Survey, "Volcanic Gases and Climate Change Overview," https://web.archive.org/web/20150812154815/http://volcanoes.usgs.gov/hazards/gas/climate.php, accessed April 25, 2016.

13. "Bachmann: 'People Don't Want Crony Capitalism,'" September 13, 2011, Fox News, http://video.foxnews.com/v/1156812920001/bachmann-people-dont-want-crony-capitalism/?#sp=show-clips.

14. "Defensive? Sen. Rand Paul on Voluntary Vaccines," CNBC, February 2, 2015, http://video.cnbc.com/gallery/?video=3000351424.

15. "CNN Reagan Library Debate: Later Debate, Full Transcript," September 16, 2015, CNN, http://cnnpressroom.blogs.cnn.com/2015/09/16/cnn-reagan-library-debate-later-debate-full-transcript.

16. Institute of Medicine, "Adverse Effects of Vaccines: Evidence and Causality," August 2011, http://www.nationalacademie.org/hmd/Reports/2011/Adverse-Effects-of-Vaccines-Evidence-and-Causality.aspx.

17. J. W. Peters, "Rand Paul Gets a Booster Vaccination," *New York Times*, February 3, 2015, http://www.nytimes.com/politics/first-draft/2015/02/03/rand-paul-gets-a-booster-vaccination/?_r=0.

18. "Rand Paul: Most Vaccines Should Be 'Voluntary,'" YouTube, February 2, 2015, https://www.youtube.com/watch?v=CGvBB_nqZWI.

19. Centers for Disease Control and Prevention, "Hepatitis B VIS," current edition date: February 2, 2012, http://www.cdc.gov/vaccines/hcp/vis/vis-statements/hep-b.html.

20. F. DeStefano, C. S. Price, and E. S. Weintraub, "Increasing Exposure to Antibody-Stimulating Proteins and Polysaccharides in Vaccines Is Not Associated with Risk of Autism," *Journal of Pediatrics* 163 (2013): 561–67, http://www.jpeds.com/article/S0022-3476%2813%2900144-3/abstract.

21. M. J. Smith and C. R. Woods, "On-Time Vaccine Receipt in the First Year Does Not Adversely Affect Neuropsychological Outcomes," *Pediatrics* 125 (2010): 1134–41, http://pediatrics.aappublications.org/content/125/6/1134.

22. S. J. Hambidge, S. R. Newcomer, K. J. Narwaney, et al., "Timely versus Delayed Early Childhood Vaccination and Seizures," *Pediatrics* 133 (2014): e1492–99, http://pediatrics.aappublications.org/content/early/2014/05/14/peds.2013-3429.

23. D. Levitan, "Paul Repeats Baseless Vaccine Claims," FactCheck .org, February 3, 2015, http://www.factcheck.org/2015/02/paul -repeats-baseless-vaccine-claims. Quotes provided to the author while employed at FactCheck.org.

24. Centers for Disease Control and Prevention, "Report Shows 20-Year US Immunization Program Spares Millions of Children from Diseases," April 24, 2014, http://www.cdc.gov/media/ releases/2014/p0424-immunization-program.html.

CONCLUSION: THE CONSPICUOUS SILENCE

1. "A Timeline of HIV/AIDS," AIDS.gov, https://www.aids.gov/ hiv-aids-basics/hiv-aids-101/aids-timeline, accessed April 25, 2016.

2. P. Boffey, "Reagan Defends Financing for AIDS," *New York Times*, September 18, 1985, http://www.nytimes.com/1985/09/18/ us/reagan-defends-financing-for-aids.html.

3. G. M. Boyd, "Reagan Urges Abstinence for Young to Avoid AIDS," *New York Times*, April 2, 1987, http://www.nytimes.com/1987/04/02/ us/reagan-urges-abstinence-for-young-to-avoid-aids.html.

4. A. White, "Reagan's AIDS Legacy/Silence Equals Death," SFGate, June 8, 2004, http://www.sfgate.com/opinion/openforum/article/ Reagan-s-AIDS-Legacy-Silence-equals-death-2751030.php.

5. "Surgeon General's Report on Acquired Immune Deficiency Syndrome," October 22, 1986, http://profiles.nlm.nih.gov/ps/access/ NNBBVN.pdf.

6. Boffey, "Reagan Defends Financing for AIDS."

7. American Foundation for Aids Research (amfAR), "Thirty Years of HIV/AIDS: Snapshots of an Epidemic," http://www.amfar.org/ thirty-years-of-hiv/aids-snapshots-of-an-epidemic, accessed April 25, 2016.

8. Science Debate, "Board & Advisory Committee," http://www .sciencedebate.org/board, accessed April 25, 2016.

9. "The Top American Science Questions: 2012", Science Debate, September 4, 2012, http://www.sciencedebate.org/2012/debate12.

10. Lawrence Krauss, e-mail to the author, December 21, 2015.

11. Shawn Lawrence Otto, e-mail to the author, December 22, 2015.

12. White House, Office of the Press Secretary, "Remarks by the President in State of the Union Address," January 20, 2015, https://

www.whitehouse.gov/the-press-office/2015/01/20/remarks-presi
dent-state-union-address-january-20-2015.

13. C. Richert, "Minnesota Republicans Change Their Tone on Climate Change," *MPR News*, June 23, 2015, http://www.mprnews
.org/story/2015/06/23/climate-change-republicans.

Index

Note: Page numbers in *italics* refer to illustrations.

abortions:
  and fetal pain, 9–16
  and fetal tissue, 90, 93
  misinformation spread about,
    88, 95, 186
  and rape cases, 186–88
  Republican Party position on,
    9–10, 11, 16
  restrictions on, 27, 93
  and women's rights, 10
Abraham, Ralph, 10, 14–15
acid rain, 2, 127, 149–50
Acorn Fork, Kentucky, natural
  gas well site spillage in, 117,
  119
aging process, research on, 103
AIDS, 201–3
air quality, 125–31
  and cap-and-trade, 131–32, 134
  and technological innovation,
    136
  see also greenhouse gases

Akin, Todd, 186, 187–88
Alaska:
  glaciers in, 38–42, *39*
  seafood industry in, 170
alcohol:
  animal studies on use of, 19
  gateway effect of, 17, 21, 22
Al Qaeda, 97
Alzheimer's disease, research on,
  108
American Coalition for Clean
  Coal Electricity, 147
American Congress of Obstetri-
  cians and Gynecologists, 12
American Fuel & Petrochemical
  Manufacturers, 147
American Geophysical Union
  (AGU), 53
American Psychological Associa-
  tion (APA), 189
American Society for Cell Biol-
  ogy, 90

ammonium persulfate, 115
amniocentesis, 11, 13
anaphylaxis, 197
Antarctic ice sheet, 45
anthropogenic forcing, 101
anti-science, 204
anti-vaccination movement, 6,
    60, 64, 65, 67, 195–200
AquaBounty, 168, 169
AquAdvantage salmon (GE) fish,
    166–70
Arizona, solar power in, 132
armadillos, disease transmitted
    via, 69
Armstrong, Neil, 49
Army Corps of Engineers:
    and wetlands protection, 183
Arndt, Deke, 25–26
arsenic, 117
asthma, 197
Atlantic, 188
autism, 63–64, 195, 196–97,
    199
avian influenza, 54–55, 56

babies:
    anencephalic, 13
    premature, 110
Bachmann, Michele, 195
barium, 117
Barnett Shale formation, North
    Texas, 117
Battelle Memorial Institute, 158,
    159–60
Bellen, Hugo, 106, 108
Bell's palsy, 197
Benishek, Dan, 9
benzene, 115

Berkeley Earth, 78
Big Hill Acres, Mississippi, 183
Biological Psychiatry, 19
biorepository, 92
biospecimen research, 92
bird flu, 54–55, 56
birth defects, testing for, 13
Blame the Blogger, 73–95
    and climate change, 74–75
    and global warming, 80–84
    and Planned Parenthood,
        88–95
    see also Internet
Blind Eye to Follow-Up,
    155–71
    and "Climategate scandal,"
        161–66
    and frontal lobotomy, 155–56,
        161, 171
    and genetic research, 156–61
Boehner, John, 4
Bolden, Charles, 50
Bonnen, Dennis H., 151, 153
Booker, Christopher, 75, 77–78,
    79
Boustany, Charles, 9
Bozeman, Barry, 160–61
BP oil well blowout, 131
brain:
    and gateway drugs, 17
    pain experienced in, 11–12,
        13–14
BRAIN Initiative, 160
British Medical Journal, 63
Brooks, Mo, 60–62, 64, 66
Buchanan, Pat, 67–68, 69
*Bulletin of the American Meteorolog-
    ical Society,* 50, 87

Burke, Marshall, 146
Bush, George W.:
  administration of, 135, 183
  and budget cuts, 56–58, 137
  on global warming, 140
  and NIH funding, 55–58, 104
  on renewable energy, 135, 137
Bush, Jeb, 141, 204
Butter-Up and Undercut, 44–59
  and NASA's funding, 44–47,
    49–50, 51–52, 54
  and NASA's mission, 44,
    46–49
  and NIH's funding, 54–59

California:
  solar power in, 132
  vaccination rates in, 64–65
  wind power in, 130
cancer:
  and genetic research, 160
  and HPV, 151–52
  pain of, 123
  precursors of, 152
cap-and-trade, 131–32, 134,
  144–45
Cape Canaveral, Florida, 50–51,
  51
carbon dioxide, 98, 128–29, 134,
  148, 193–94
Carson, Ben:
  on immigration, 66
  on sexual orientation, 188–89,
    191–92
Carter, Jimmy, 1
Cato Institute, 177–79
Center for Bio-Ethical Reform,
  94

Center for Medical Progress
  (CMP), 88, 91, 93, 94
Centers for Disease Control and
  Prevention (CDC):
  favorability rating of, 45
  on immigrants and disease, 62
  and vaccination, 61, 153, 197–
    200
  on water contamination, 115
Certain Uncertainty, 138–54
  in climate science, 139–50
  and HPV vaccine, 151–54
cervical intraepithelial neoplasia
  (CIN), 152
Chagas fever, 68, 69
Cherry, James, 200
Cherry-Pick, 28–43
  and climate change, 28–43
  and TOADS, 29–31
chlorofluorocarbons, 98
Christie, Chris:
  on cap-and-trade, 131–32,
    134
  on marijuana, 16–17, 19, 21,
    22
  and solar power, 132–34
Christy, John, 35–36
chub, creek, 117
CitiGroup, 148
civil rights:
  and marriage equality, 188–
    89, 192
  and sexual orientation, 188–
    92
Clean Air Act, amendments to
  (1990), 127
Clean Air Interstate Rule (2005),
  127

Clean Power Plan, 134, 146–47, 148

Clean Water Act (1972), 181, 185

Clean Water Rule, 180–81

climate:
  human influences on, 101, *143*
  long-term, 30, 33, 36, 37

Climate Action Report (US State Dept.), 129

climate change:
  and Alaska's glaciers, 38–42, *39*
  and Blame the Blogger, 74–75, 80–84
  and Certain Uncertainty, 139–50
  and Cherry-Pick, 28–43
  deniers of, 26, 27, 28–31, 45, 80–84, 162–63, 194
  and disease, 68–69
  economic effects of, 4, 145–49
  global agreement on, 148
  and global temperatures, 23–27, 30, 74–79
  and greenhouse gases, 128
  Internet blog sources on, 74–75, 80–84
  misinformation spread about, 3–4, 6, 165
  Oversimplification, 23–27
  Pentagon on, 99–101
  preventing action on, 165–66
  and public health, 147
  Ridicule and Dismiss, 96–101, 102
  and rising sea levels, 4, 50–51, 99–100, 139, 147, 205
  scientific consensus on, 81
  and terrorism, 100, 101
  and TOADS, 29–31, 74, 75, 80, 88, 164, 194
  and warming hiatus, 31, 33–34
  see also global warming

"Climate Change Adaptation Roadmap" (Pentagon), 99–101

Climate Dynamics, 144

"Climategate scandal," 161–66

climate science:
  evolution of, 86, 88, 149, 206
  fingerprint analysis in, 142–43, *143*
  NASA's activities in, 45–47, 81, *143*, 205
  paleoclimatology, 139

Climatic Research Unit (CRU), East Anglia, 161–64

climatology, *see* climate science

Clinton, Bill, 69

coal:
  burning, 129
  and climate change, 149

Coburn, Tom, 101–2, 109

cocaine:
  tolerance for, 19
  use of, 18, 22

Collins, Francis, 58

Columbia University, 50, 51

Commerce Department, US, 163

Compton, Carolyn, 91–92

Conspicuous Silence, 201–6
  and AIDS, 201–2
  on HIV, 202–3
  and science debates, 203–4

correlation vs. causation, 18, 63–64, 115–16, 137

cortex, 12, 13, 14

Couric, Katie, 193, 194

Credit Snatch, 124–37
  and air quality, 125–31
  and cap-and-trade, 131–32, 134
  and elections, 135, 137
  and solar power, 132–34, 135–37

Cruz, Ted:
  on global cooling, 85–86, 87, 88
  on global warming, 31, 33–34, 37
  on hard vs. soft sciences, 52–54
  on NASA's funding, 44–47, 51–52, 54
  and NASA's mission, 44, 47, 49

Cuomo, Chris, 189

D68 virus, 62

dace, blackside, 117, 119

Daleiden, David, 88, 89, 94

Dear, Robert, 95

decadal surveys, 53–54

Democratic Party, on women's rights, 10

Demonizer, 60–72
  and immigration, 60–63, 66–72
  and vaccinations, 60, 64–66

dengue fever, 68–69

Development and Psychopathology, 20

diabetes, and vaccines, 197

Diesel Emissions Reduction Incentive, 126, 128

diphtheria, 197

Disneyland, 60, 64, 65

diurnal drift, 36

Drosophila melanogaster, see fruit flies

Drug Enforcement Administration (DEA), 120–21

DTaP vaccine, 197

Ebola outbreak (2014), 58

education, STEM (science, technology, engineering, and math), 44–45

Ellis Island, 67

El Niño, 32

Emmerich, Wyatt, 182, 183, 184

Enders, John Franklin, 91

Energy Department, US:
  budget cuts of, 137
  and SunShot, 136–37
  on wind power, *136*

Energy Information Administration (EIA), 127, 128, 130–31

Energy Policy Act (2005), 137

enterovirus, 61–62

Environmental Protection Agency (EPA):
  and Clean Power Plan, 134, 146–47, 148
  and Clean Water Rule/Act, 180–81, 185
  and "Climategate scandal," 163–64

Environmental Protection
Agency (EPA) (*continued*)
and fish, 172–79
and fracking, 115, 117
and greenhouse gas emissions,
129
and Mercury and Air Toxics
Standard, 173–74
and Mount St. Helens, 2
and wetlands protection, 182–85
Environmental Science & Technology, 117
epigenetic effects, 190
European Neuropsychopharmacology, 19
Exit Glacier, Alaska, 39–40, *39*

Fabius Maximus (FM) blog,
81–84
fabrications, *see* Straight-Up Fabrication
Face the Nation, 144
FactCheck.org, 91
factories, as point sources, 2,
126, 127
fallacy of anecdotal evidence, 29,
42–43, 118, 154
fecundity, 191
Fergusson, David, 21–22
fertility, 191
fetal pain, 9–16
fetal tissue, 88–94
Fiorina, Carly, 89, 90, 93–94
fish, 172–79
genetically engineered (GE),
166–70
"high-end" subsistence fishers,
178–79

mercury in, 173–76, 178–79
recreational angler households,
174, 176–77, 179
Florida, and rising sea levels,
50–51
fluorinated gases, 99
flu vaccines, 197
flu viruses, 54–55
Food and Drug Administration
(FDA), and GMOs, 166–69
fossil fuels:
burning, 2, 48, 129–30, 146
oil prices, 137
Fox, Josh, 114
fracking, 112–20
cement integrity of wells in,
114, 117, 119
process of, *113,* 118, 119
regulation of, 114, 119
and water contamination, 112,
114–20, 122
fracking fluid, 113–15, 117,
118–19
Frankenfish, 166–70
frontal lobotomy, 155–56, 161, 171
fruit flies, *105*
as model organism, 103, 107, 108
research on, 101–9
fuel switching, 129

Gardasil, 151–53, 195
Gasland (documentary film), 114,
120
gateway drugs, 16–22
biological mechanisms underlying, 21
effects on body and brain, 17
twin studies on, 19–20

General Accounting Office, 92–93

genes, and heredity, 107

genetically engineered (GE) fish, 166–70

genetic manipulation:
  and fruit fly research, 108, 109
  GMOs, 166–70

genetic mutations, 107–8

genetic research, 106–9
  and cancer, 160
  Human Genome Project, 156–61

genetic testing, 13

George C. Marshall Institute, 35

geothermal energy, 137

Gibson, Lauren, 191–92

Gingrich, Newt, 105

GISS, 36

glaciers:
  catchment basins of, 41
  ice loss from, 41
  mass imbalance in, 40
  receding or growing, 38–42, 39
  "reference," 42

global cooling:
  as red herring, 4
  sources of myth about, 85–88
  and sulfates in stratosphere, 194

Global Historical Climatology Network, 79

global temperatures, 23–27, 74–79
  and cherry-picking data, 32–43
  corrections and adjustments to, 34–35, 77, 78, 79

and El Niño, 32

fingerprint analysis of, 142–43, 143

human impact on, 143, 144, 205

measurement of, 25, 48, 75–76, 77, 139

regional variations in, 33, 40, 76

satellite data on, 34–37

global warming:
  and Blame the Blogger, 80–84
  and Certain Uncertainty, 139–50
  and Cherry-Pick, 29, 31
  and "Climategate scandal," 162–66
  and greenhouse gases, 98
  human causes of, 140–45
  and Straight-Up Fabrication, 194
  as threat multiplier, 99
  and TOADS, 29–31
  and warming hiatus, 31, 33–34
  see also climate change; global temperatures

GMOs (genetically modified organisms), 166–70
  evidence of no harm in, 166–67, 168
  purposes of, 166

Golden Fleece Awards (Proxmire), 109

Golden Goose Awards, 109–10

GOP, see Republican Party

Graham, Lindsey, 7

Grantham Collection, 94
greenhouse gases, 98, 125–29,
    144
  and climate change, 128
  and global warming, 98
  lawsuits involved in, 130–31
  Regional Greenhouse Gas Ini-
    tiative, 131–32, 134
Greenland ice sheet, 42, 45
Greenwood (Mississippi) *Com-*
    monwealth, 182
groundwater contamination, 114
Grueber, Martin, 158
Gwynne, Peter, 86

HAART (highly active anti-
    retroviral therapy), 202–3
HADCRUT, 36
Hansen's disease, 69
health care:
  and immigration, 71–72
  savings in, 110, 147
Helms, Jesse, 70
Hensley, Christopher, 191–92
hepatitis, 68, 198
herd immunity, 65
heroin, 18, 121
HIV, 69–70, 202–3
HIV-associated sensory neurop-
    athy, 121–22
Homewood, Paul, 74–75, 77–78
homosexuality, 190–92
  and AIDS, 201–2
  and choice, 189–91, 200
  discrimination and harassment
    of, 192
  and fecundity, 191
  genetic component in, 190–91

horizontal drilling, 113, 116
HPV (human papillomavirus)
    vaccine, 151–54, 195–97
Hubbard Glacier, Alaska, 39
Huckabee, Mike, on climate
    change, 96–101, 102,
    193–94
Huffman, Jared, 167, 169
Human Genome Project,
    156–61
hydraulic fracturing, *see* frack-
    ing
hydrochloric acid, 115
hydrochlorofluorocarbons, 98

ice age, 4, 85
ice thickness, measurement of,
    34, 139
immigration, 60–63
  and disease, 60–62, 64,
    66–72
  disputes over resources and,
    100, 205
  and health care, 71–72
  and vaccinations, 65–66
Immigration Act (1917), 70
industrial facilities:
  and acid rain, 150
  as point sources, 2, 127, 128–
    29
Inhofe, James:
  Cherry-Pick by, 28–29, 30
  and "Climategate scandal,"
    164–66
  The Greatest Hoax, 164
  Literal Nitpick by, 112, 113–
    14, 117–20, 122
Inside Science, 86

Institute of Medicine (IOM), 17–18, 197
Intergovernmental Panel on Climate Change (IPCC), 81–84
on causes of climate change, 144
climate goals set, 165
Fifth Assessment Report of, 142
Interior Department, US:
and fracking, 114
on wetlands, 184–85
Internet:
on climate change, 74–75, 80–84
false information spread via, 64, 73–74, 80, 81, 89, 93, 95
on global cooling, 85–88
level playing field of, 73, 85
Planned Parenthood videos on, 88–95
Iowa Ag Summit (2015), 131
IQ deficits, 173, 175, 178
IRS (Internal Revenue Service), 45
ISIS:
beheadings by, 99, 100
spread of, 97

Jackson (Mississippi) *Northside Sun,* 182
Jacobson, Mark Z., 146
JAMA (*Journal of the American Medical Association*), 13–14, 19
Jeffries, Michael, 188
Jindal, Bobby, 7
Journal of Geophysical Research, 78–79

Journal of Pediatrics, 199
Jurassic Park (film), 169
Justice Department, US, 183, 184

Kennedy, John F., 48
Kennedy Space Center, Florida, 50–51, *51*
kissing bug, 69
Krauss, Lawrence, 204

Lancet, 63
leprosy, 68, 69
Lewandowsky, Stephan, 140–41
LGBTQ community, 192
Literal Nitpick, 111–24
and fracking, 112–20, 122
and medical marijuana, 120–23
Lost in Translation, 172–85
and EPA, 172–84
and fish, 172–79
and water quality, 180–81, 185
and wetlands, 182–85
LSD, 121
Lucas, Robert, 180, 181–84
Lynch, Loretta, 121

Maher, Bill, 80, 81, 84
malaria, 68
manufacturing, decline in US, 128–29
marijuana:
as gateway drug, 16–22
medical uses of, 120–23
Schedule 1 status of, 121
THC in, 19, 121

Marijuana and Medicine (Institute
    of Medicine), 17–18
Markel, Howard, 71
marriage equality, 188–89, 192
MATS rule (Mercury and Air
    Toxics Standard), 173–74
Mayo Clinic website, 73
McComb (Mississippi)
    Enterprise-Journal, 182, 183
McConnell, Mitch, 4, 85, 88
McEntee, Christine, 53
McKenna, Mike, 5
Mears, Carl, 36–37
measles, 60–61, 62–65, 91
Merck Pharmaceutical, 152
Mercury and Air Toxics Stan-
    dard (MATS rule), 173–74
mercury emissions, 173–76,
    178–79
Merker, Bjorn, 13
methamphetamines, use of, 18
methane, 98, 116, 119
methylmercury, 176
Middle East, radical groups in,
    96, 97, 99
Mississippi Department of
    Health, 183
MMR vaccine, 63–64, 195–97,
    200
model organisms, 103, 107, 108
Moniz, Egas, 156, 161
Morgan, Thomas Hunt, 107, 108
mosquitoes, 68, 69
Mount Pinatubo, Philippines,
    193–94
Mount St. Helens eruption, 1–3,
    194
Müller, Hermann, 107–8

Muller, Richard, 78
multiple sclerosis, 122
mumps, 63–64
Murkowski, Lisa, 167, 168–69
Murphy, Matt, 60, 74

naphthalene, 115
NASA (National Aeronautics
    and Space Administration):
    and climate change, 45–47, 81,
        143, 205
    creation of, 47–48
    funding of, 44–50, 51–52, 54
    on global temperatures,
        23–26, 33, 34, 36, 48, 78
    on human influence on cli-
        mate, *143*
    mission of, 44, 46–49
    and rising sea levels, 50–51, *51*
National Academy of Sciences,
    17–18, 53, 101, 116, 150
National Aeronautics and Space
    Act (1958), 47–48
National Biomarker Develop-
    ment Alliance, 92
National Cancer Institute, 92
National Economic Research
    Associates (NERA), 146–47,
    148
National Institute on Drug
    Abuse (NIDA), 17, 19
National Institutes of Health
    (NIH):
    basic scientific research in,
        56–59
    and Ebola outbreak, 58
    funding of, 55–59, 102, 104–5
    Revitalization Act (1993), 91

National Mining Association, 147

National Research Council, 87

National Science Foundation (NSF), 57, 104

National Snow and Ice Data Center, 42

National Survey of Fishing, Hunting, and Wildlife-Associated Recreation, 176

natural gas:
  as fossil fuel, 129
  fracking for, 112–20

natural gas boom, 122–23, 128

Nature, 146, 160

Nature Geoscience, 144

Netherlands, drug use in, 22

Netherlands Environmental Assessment Agency, 81–84

neuroanatomy, 11, 12

neurobiology, 108

Neurology, 121

Neuropsychopharmacology, 19

neuroscience, 107

New Jersey:
  and Regional Greenhouse Gas Initiative, 131–32, 134
  solar power in, 132–34
  Solar Renewable Energy Certificate (SREC) program in, 133–34

Newsweek, 86, 88

New York Times, 105

Nickles, Don, 69, 70

nicotine:
  animal studies on use of, 19
  gateway effect of, 17, 21, 22

nitrogen oxide, 125, 126

NOAA (National Oceanic and Atmospheric Administration), 205
  and Global Historical Climatology Network, 79
  on global temperatures, 23–27, 33, 34, 36, 78
  ocean-based climate station of, 76

nociception, 14

North Dakota, fracking in, 113

"not a scientist":
  as dumb talking point, 5
  Obama's statement, 205
  Reagan's statement, 1, 3, 127, 192
  as Republican talking point, 7

NOVA website, 73

Nucatola, Deborah, 90, 91

Obama, Barack:
  in Alaska, 38–40
  and BRAIN Initiative, 160
  and Clean Power Plan, 146
  on climate change, 23–27, 96–98, 205
  and disease prevention, 70
  and Human Genome Project, 156–57, 159, 161
  and NIH funding, 56
  on radical groups in Middle East, 96

oil and gas industry:
  and cap-and-trade, 131–32, 134
  and climate change, 149
  and fracking, 112–20

oil and gas industry (*continued*)
  offshore drilling, 131
  tax breaks for, 137
Oklahoma, fracking in, 118
olfaction, research on, 103
Otto, Shawn Lawrence, 204
Oversimplification, 9–27
  and abortions, 9–16, 27
  on climate change, 23–27
  on gateway drugs, 16–22, 27
Oxburgh Report, 163
ozone hole, 98
ozone levels, 125

pain:
  and the brain, 11–12, 13–14
  fetal, 9–16
  and medical marijuana, 122, 123
  as subjective experience, 10
Pain-Capable Unborn Child Protection Act (2015), 9
paleoclimatology, 139
Palin, Sarah, 38–39, 41, 42
Palmer, Gary:
  on climate change, 74, 79, 80
  and "Climategate scandal," 162–63, 164, 166
pandemics, 54, 55, 100
Paul, Rand:
  on fruit fly research, 102–5, 107, 108–9
  Government Bullies, 181, 184
  and Lucas story, 180, 181–82, 183

and Planned Parenthood, 89, 90
  on vaccinations, 195–96, 198, 200
Pediatrics, 199–200
Pennsylvania:
  fracking in, 113
  oil and gas development in, 116
  water supply safety in, 116
Pentagon, on climate change, 99–101
Perry, Rick:
  on greenhouse gas emissions, 125–29
  on HPV vaccine, 151–53
  and Planned Parenthood, 89, 90
  as Texas governor, 130–31
pertussis, 197
petroleum distillate, 115
Pew Research Center, 45
phenotypes, 106
pheromones, research on, 104
Pierrehumbert, Raymond, 35
Planned Parenthood, 88–95
anti-abortion video about, 88
  fetal tissue sales misrepresented, 88–94
  violence against, 88, 94–95
Pletcher, Scott, 103–4, 105
polio, 68, 91
politicians:
  credit claimed by, 124–25, 130–31, 137
  "do nothing" urged by, 138–39, 140–41, 144, 146, 148–50, 165

and elections, 135, 137
ideology promoted by, 153, 165, 171
misinformation spread by, 4–6, 93, 95, 187, 206
public influence of, 6, 95
scientific research belittled by, 102–5, 108–9, 110
words chosen by, 14, 112, 118–20, 123, 153
pollution:
  and cap-and-trade, 131–32, 134
  and credit snatch, 125–34
  from point sources, 2, 126, 127–29
Potts, Brian, 177, 179
power plants:
  and acid rain, 150
  reducing emissions from, 127–28, 130
precipitation patterns, changing, 99–100
precipitation rates, 40–41, 139
Precision Medicine Initiative, 156–57
pregnancy, 187
presidency:
  expert knowledge available to, 1–3, 205
  political opponents' belittling of, 101
prison, and sexual orientation, 189, 191–92, 200
Proxmire, William, 109
psychosurgery, 156, 171

quantum physics, 173

Racketeer Influenced and Corrupt Organizations Act (RICO), 180, 181
rape:
  and abortion, 186–88
  in college culture, 187–88
  "legitimate," 186–87, 200
rat pups, massage of, 110
Reagan, Ronald:
  and acid rain, 2, 149–50
  and AIDS, 201–3
  on Mount St. Helens eruption, 1–3, 194
RealClimate blog, 35
Regional Greenhouse Gas Initiative, 131–32, 134
Remote Sensing Systems, 36
Republican Party:
  on abortions, 9–10, 11, 16
  antiscience attitudes in, 7, 105
  and Certain Uncertainty, 140
  climate deniers in, 28–29, 45
  and fracking, 114
  government spending opposed by, 102, 104, 108
  misinformation spread by, 102
  "not a scientist" as mantra of, 7
  political agendas spread by, 93, 104
  as the stupid party, 7
Revkin, Andy, 25
Ridicule and Dismiss, 96–110
  and climate change, 96–101, 102
  fruit fly research, 101–8
  and "Wastebook," 101–2, 109

Robbins, Frederick Chapman,
91
Romney, Mitt, 203–4
Rosenberg, Chuck, 120–21,
122
Royal College of Obstetricians
and Gynaecologists (UK),
12
rubella, 63–64, 91
Rubio, Marco:
and Certain Uncertainty,
144–45, 146, 149
on climate change, 3, 144–45,
146

salmon, *169*
farm-raised, 168
genetically engineered (GE),
166–70
wild populations of, 166–69
Santorum, Rick:
and fish, 172–78, 179
and global warming, 80–81,
83, 84
Sawyer, Sherilyn, 92
Science, 33–34, 104, 190
Science Debate, 204
Science.gov, 73
scientific issues:
antiscience attitudes about, 7,
105, 108–10, 204
correlation vs. causation in,
18, 63–64, 115–16, 137
evolution of, 6–7, 59, 156,
170–71
funding research on, 56–59,
102, 104–5, 161
hard vs. soft, 52–54

major discoveries in, 110
margin for error in, 154
raw data with differing outputs in, 36, 43
return on investment in, 157–
58, 160–61
as sound bites, 11, 15, 27, 83, 110
uncertainty in, 17, 20, 22,
26–27, 36–37, 154
see also specific issues
Scott, Rick, on climate change,
3–4
seafood industry, 166–70
sea levels, rising, 4, 50–51, *51,*
99–100, 139, 147, 205
selenium, 117
service-based economy, US shift
to, 129
sex-linked transmissions, 190
sexual orientation, 188–92
and choice, 189–91, 200
discrimination and harassment, 192
genetic component in, 190, 191
sexual violence, 188
Sierra Club, 37
social media, impact of, 6, 206
solar activity, 143, 144
Solar Energy Industries Association, 132
solar power, 132–34, 135–37
Space Act (revised 1984), 49
Spanish flu, 54
Spencer, Roy, 35–36
Sputnik, 48
Stanford University, 146
Starkville (Mississippi) *Daily
News,* 182

State Department, US, *Climate Action Report,* 129
State of the Union (CNN), 38
STEM (science, technology, engineering, and math) education, 44–45
Stern, Alexandra Minna, 71
Straight-Up Fabrication, 186–200
  on abortion in rape cases, 186–88
  and anti-vaccination movement, 195–200
  on sexual orientation, 188–92
  on volcanic emissions, 193–94
strontium, 117
sulfate aerosols, 143, 194
sulfur dioxide, 1–3, 125, 126–27, 150, 193
sunburn, 97–99, 101
SunShot, 136–37
Supreme Court, US, Cato Institute's brief to, 178–79
swine flu, 54
Syria:
  civil war in, 100–101
  drought in, 101

tanning beds, 97
Telegraph (London), 75
terrorism:
  and climate change, 100, 101
  domestic, 95
tetanus, 197
Texas:
  HPV vaccine in, 151–53
  oil and gas interests in, 131
  Renewal Portfolio Standard in, 130
  wind power in, 129–30

thalamus, 11
THC (tetrahydrocannabinol), 19, 121
Thomas, Lee, 149–50
Thompson, M. E. Jr., 182–83
Thorazine, 171
Time, 87, 88
TOADS:
  and Blame the Blogger, 74, 75, 80, 88
  and Blind Eye, 164
  and Cherry Pick, 29–31
  and climate change, 29–31, 74, 75, 80, 88, 164, 194
  and Straight-Up Fabrication, 194
travel, and disease, 68
triethanolamine zirconate, 115
Tripp, Simon, 158
troposphere, temperatures recorded in, 34
Trump, Donald:
  anti-vaccination views presented by, 196
  border wall proposed by, 67
tuberculosis, 68, 70–71
twins, genetic code shared by, 20
twin studies:
  on gateway drugs, 19–20
typhoid fever, 70

ultraviolet (UV) radiation, 97–98
uncertainty, *see* Certain Uncertainty
University of Alabama, Huntsville (UAH), 35–36

University of Colorado, Boulder, 42
University of East Anglia, 161–63
University of Michigan, 71
University of Otago, New Zealand, 21
University of Tennessee, Chattanooga, 191–92
University of Washington, 36
US Fish and Wildlife Service, 117
US Geological Survey, 2, 117, 193–94

vaccinations:
  and autism, 63–64, 195, 196–97, 199
  CDC on, 61, 153, 197–200
  DTaP, 197
  and herd immunity, 65
  HPV, 151–54, 195–97
  international rates of, 65–66, 72
  lives saved via, 200
  measles, 60–61, 62–65, 91
  and mental disorders, 195–96, 197
  misinformation about, 6, 63–64
  MMR vaccine, 63–64, 195–97, 200
  opposition to, 6, 60, 64–67, 195–200
  personal belief exemption from, 64–65
  safety of, 196–97, 199
vehicles, as mobile sources of pollution, 126
volcanic eruptions, 1–3, 143, 192–94
voting public, 206

Wakefield, Andrew, 63
Wall Street Journal, 177, 179
"Wastebook" (Coburn), 101–2, 109
water contamination:
  Clean Water Rule/Act, 180–81, 185
  and dumping in wetlands, 181–84
  and fracking, 112, 114–20, 122
"waters of the United States," definition of, 181
weather:
  extreme events, 99–100, 139
Weiss, Susan, 17, 19, 20
Weller, Thomas Huckle, 91
wetlands:
  definition of, 182, 184
  disappearance of, 184
  importance of, 184–85
  protection of, 182–85
White House Office of Science and Technology, 150
whooping cough, 197
Will, George, 4
wind power, 129–30, 135–37, 136
women, and control over one's own body, 10
World Glacier Monitoring Service, 42
World Health Organization (WHO), 64, 71, 170
Wrigley, Robbie Lucas, 182–83

X-rays, 107

Young, Don, 167, 168, 169
YouTube, 73